AGRICULTURE AND THE ENVIRONMENT

AGRICULTURE AND THE ENVIRONMENT

NOEL URI

Nova Science Publishers, Inc.
Commack, New York

Editorial Production: Susan Boriotti
Office Manager: Annette Hellinger
Graphics: Frank Grucci and John T'Lustachowski
Information Editor: Tatiana Shohov
Book Production: Donna Dennis, Patrick Davin, Christine Mathosian
and Tammy Sauter
Circulation: Maryanne Schmidt
Marketing/Sales: Cathy DeGregory

Library of Congress Cataloging-in-Publication Data
available upon request

ISBN 1-56072-654-7

6080 Jericho Turnpike, Suite 207
Commack, New York 11725
Tele. 516-499-3103 Fax 516-499-3146
e-mail: Novascience@earthlink.net
e-mail: Novascil@aol.com
Web Site: http://www.nexusworld.com/nova

Printed in the United States of America

TO CAROL

CONTENTS

Preface		**ix**
Chapter 1	Introduction and Overview	**1**
Chapter 2	Agricultural Chemical Use	**7**
Chapter 3	The Economic and Environmental Implications of Alternative Production Practices	**41**
Chapter 4	The Influence of Government Policy on the Choice of Production Practices and Chemical Use	**59**
Chapter 5	Government Policy and the Development and Use of Biopesticides	**115**
Chapter 6	The Changing Pattern of Soil Erosion in the United States	**139**
Chapter 7	Perceptions versus Reality: The Case of No Till Farming	**161**
Chapter 8	The Use of Conservation Practices in Agriculture in the United States	**175**
Chapter 9	The Impact of Energy on the Adoption of Conservation Tillage in the United States	**211**
Chapter 10	The Environmental Consequences of the Conservation Tillage Adoption Decision in Agriculture in the United States	**237**
Index		**271**

PREFACE

Agriculture in the United States is in the midst of a major transition motivated by economic and environmental factors. These include water quality and quantity, soil erosion, and the compatibility of agricultural production practices and the quality of the environment. In the context of this change, U.S. agricultural policy seeks to balance several objectives including an abundance of food and fiber at reasonable prices, economic security for agricultural producers, and conservation of natural resources.

Agricultural chemical use and soil and water quality degradation associated with agricultural production are significant among the environmental problems confronting the United States. In fact, these are now perceived as environmental problems comparable to other environmental problems such as air quality deterioration and the release of toxic pollutants from industrial sources. While the growth of agricultural chemical use is an integral part of the technological revolution in agriculture that has generated major changes in production techniques, uncertainties about the health effects of agricultural chemicals are very important concerns. Severe soil degradation from erosion, compaction, or salinization can destroy the productive capacity of the soil. It can also impair water quality from sediment and agricultural chemicals.

This book looks at both of these significant issues - the relationship between agricultural chemical use and the environment and the relationship between soil and water quality degradation associated with agricultural production and the environment. After an overview provided in the first chapter, the next four chapters focus on agricultural chemical use. The subsequent chapters look at soil erosion, water quality issues, and agricultural production. Due to the nature of policy issues involved, however, the material in each chapter is inexorably intertwined with the material in other chapters. Consequently, cross-referencing is essential and each individual chapter must be considered in the context of the entire book.

There are a number of themes that run throughout. First, agricultural producers do respond to economic factors. These can take a number of forms including market price signals and government subsidies. Second, the government can influence the choice of production practices through a variety of means including regulation, education and technical assistance, research and development, taxes, and subsidies. Third, government intervention is warranted in some instances due to the presence of externalities. Finally, agriculture in the United States is dynamic, continually adopting new production practices and technologies. It is through a combination of these factors that the negative impacts of agriculture on the environment can be mitigated.

CHAPTER 1

INTRODUCTION AND OVERVIEW

In recent years there has been growing concern over surface water and groundwater quality, mounting public apprehension over the ecological and human health effects of agricultural chemicals, declining soil quality, and spreading resistance to pesticides. These developments have highlighted the fragility of the natural resource base of agriculture in the United States and, hence, the importance of understanding the impacts of government policies and shifts in production practices associated with technological changes on this resource base. There is a tremendous need for developing and implementing effective, socially desirable approaches for managing the resources. Additionally, current trends suggest that the importance of these issues will continue to grow as resource scarcities and environmental concerns come to play an increasingly critical role in policy debates relating to agriculture.

These considerations suggest that government agencies primarily concerned with agriculture will need to enhance their capacity to address agricultural resource management issues. This need is becoming acute as other public and private entities concerned with agriculture, the environment, and food safety are increasingly active in this area. It is obvious that there is a need to improve policy-making analysis with respect to environmental and resource management. The problems in agriculture are shared with other sectors of the economy because agriculture is a major user of land, water, and agrichemicals, and these factors affect environmental quality, recreation, wildlife, urban water supplies, and human health. Consequently, agricultural problems are a very important part of overall environmental and resource issues.

The ability to improve both the understanding and the practical management of the natural resource base of agriculture as it relates to the environment rests on three factors. These include (1) an adequate data base that can be used to define problems, describe how they have changed over time, and indicate what incentives have resulted in change, (2) a coherent body of knowledge synthesizing the findings of the research undertaken, and (3) an analytical framework for addressing these issues.

Each of these factors is used to address two important concerns involving the relationship between agriculture and the environment. These include the connection between agricultural chemical use and the environment and the association between soil and water quality degradation associated with agricultural production and the environment.

There has been a significant increase in agricultural chemical use. This has occurred in part because of the substitution among inputs and the responsiveness of input use to changes in relative prices. Beginning in the early 1970s, pesticides, for example, were becoming relatively less costly and being substituted for machinery, energy, land, and labor. Since pesticide prices represent the explicit cost to the farmer in using these inputs but ignore the costs associated with any externalities associated with their use, farmers had a strong incentive to reduce total cost and increase returns by substituting pesticides for other inputs. While relative prices have mattered in the agricultural chemical use decision, market prices for agricultural chemicals have failed to reflect the externalities. Adjustments to market prices can be made. The mechanics of such adjustments will be discussed.

Severe soil degradation from erosion can destroy the productive capacity of the soil. It can also impair water quality from sediment and agricultural chemicals. Three related causes of water quality impairment are sedimentation, eutrophication, and pesticide contamination. When soil particles and agricultural chemicals wash off a field, they may be carried in runoff until discharged into a water body or stream. Not all agricultural constituents that are transported from a field reach water systems, but a significant portion does, especially dissolved chemicals and the more chemically active, finer soil particles. Once agricultural pollutants enter a water system, they lower water quality and can impose economic losses on water users. These offsite impacts can be substantial. The offsite impacts of erosion are potentially greater than the onsite productivity effects in the aggregate. Therefore, society may have a larger incentive for reducing erosion than producers have. As in the case of agricultural chemicals, it is possible to account for the externalities associated with soil erosion.

OVERVIEW

The following chapters are logically divided into two separate parts. Chapter 2 through Chapter 5 focus on chemical use by agriculture in the United States. Chapter 5 through Chapter 10 look at soil erosion, water quality issues, and agricultural production.

Chapter 2 recognizes that agricultural chemicals are essential inputs in production agriculture in the United States. The use of agricultural chemicals has followed a varied path in response to a number of economic and technical factors. The price of output, the availability and price of the agricultural chemicals, technological constraints, and government regulation all have had an impact.

Chapter 3 notes that voluntary adoption of alternative production practices that depart from traditional approaches to production agriculture have been advanced as a way to mitigate the use of agricultural chemicals. Two general practices are discussed including crop rotations that reduce weed, disease and insect infestation problems and conservation tillage. While the adoption of either will have beneficial environmental consequences, the economic consequences are somewhat obfuscated.

Chapter 4 discusses a variety of government policies that are used to address externalities generated by agricultural production. For non-point source pollution, where the usual 'polluter pays' principle does not currently hold, penalizing agricultural producers who pollute is not feasible in the majority of cases. Government must resort to other means to encourage the adoption of alternative production practices. Since crop farmers manage about one-quarter the land mass of the contiguous United States, a voluntary policy encouraged by technical and financial assistance may be most feasible. Such a policy, however, may be ineffective in reducing agricultural chemical risks. At the other extreme, government may ban the use of a dangerous chemical, even when in some areas, due to specific soil and land characteristics, it may be relatively safely used. Banning a chemical does not guarantee that a less risky substitute will be used. Some States have resorted to taxing chemical use and also banning specific chemical use in relatively small areas where watersheds are particularly vulnerable to contamination. Indications are that taxes are ineffective means to reduce use, though they may raise revenues used for pollution abatement and clean-up. It appears that no one approach can be used successfully nationwide to achieve pesticides or fertilizer use reduction, but that a mixture of policies is appropriate.

Chapter 5 provides an assessment of the development and use of biopesticides in the future. These will emerge against the backdrop of the environmental effects associated with the use of conventional pesticides and government policies designed to control these effects. In the final analysis, farmers' choices on pesticides will be influenced by the prevailing costs and benefits of conventional pesticides and their alternatives including biopesticides. The outlook for pesticide use is complicated, though some directions can be perceived. There are a number of factors that will serve potentially to impact pesticide use which in turn will affect the development of biopesticides. These include pesticide regulation, the Federal Agriculture Improvement and Reform Act (FAIR) of 1996, the crops planted, the management of ecologically-based systems, and consumer demand for "green" products.

Chapter 6 notes that soil erosion has both on-farm and off-farm impacts. Reduction of soil depth can impair the land's productivity, and the transport of sediments can degrade streams, lakes, and estuaries. To address this problem, soil conservation policies have existed in the United States for over 60 years. Initially, these policies focused on the on-farm benefits of keeping soil on the land and increasing net farm income. Beginning in the 1980s, however, policy goals increasingly included reductions in off-site impacts of erosion. The Food Security Act of 1985 was the first major legislation explicitly to tie eligibility to receive agricultural program payments to conservation performance. The FAIR modifies the conservation compliance provisions by providing farmers with greater flexibility in developing and implementing conservation plans. As a consequence of conservation efforts, total soil erosion between 1982 and 1997 was reduced by 42 percent and the erosion rate fell from 8.0 tons per acre per year in 1982 to 5.2 tons per acre per year in 1997. Still, soil erosion is imposing substantial social costs. In 1997 these costs are estimated to have been about $29,700,000,000. To further reduce soil erosion and thereby mitigate its social costs, there are a number of policy options available to induce farmers to adopt conservation practices including education and technical assistance, financial assistance, research and development, land retirement, and regulation and taxes.

Chapter 7 recounts a number of economic and environmental benefits associated with the use of no till in production agriculture in the United States. There are lower labor, energy, and machinery costs associated with no till farming relative to conventional tillage systems and other types of conservation tillage. The reduced erosion associated with no till also leads to a number of environmental benefits including a reduction in water quality impairment. In order to fully account for the benefits associated with no till, it is important that farmers' perception of what constitutes no till and their actual use of no till be consistent. An analysis of the Agricultural Resource Management Study survey data for 1996 shows that for soybeans, winter wheat, spring wheat, and durum wheat, farmers' perceptions are consistent with reality. In the case of corn, however, nearly 18 percent of corn farmers believe they are using no till while in actuality, only slightly more than 12 percent are using this tillage system.

Chapter 8 discusses soil degradation from erosion and the attendant deterioration in water quality associated with agricultural production are significant problems. The use of appropriate conservation practices can mitigate these problems. The government has been promoting, through voluntary means, the use of such practices. There are impediments to their adoption, however, including a lack of information, a high opportunity cost associated with obtaining information, complexity of the production system, a short planning horizon, inadequate management skills, and a limited, inaccessible, or unavailable support system. Survey data for 1996 for corn, soybean, and winter wheat production show that conservation practices are still not widely adopted, although they are more prevalent on highly erodible land used in the production of corn and soybeans than on non-highly erodible land. Winter wheat remains an anomaly. There are a number of things that can be done to promote further the increased use of conservation practices. Among the public policies that can be used to affect the choices of conservation practices are education and technical assistance, financial assistance, research and development, land retirement, and regulation and taxes.

The penultimate chapter, Chapter 9, assesses the impact of energy on the adoption of conservation tillage and the special importance of energy in addressing concerns about the effect of agricultural production on the environment in the United States. After establishing that a relationship exists between the price of energy and the adoption of conservation tillage via cointegration techniques, the relationship is quantified. It is shown that while the real price of crude oil, the proxy used for the price of energy, does not affect the rate of adoption of conservation tillage, it does impact the extent to which it is used. Finally, there is no structural instability in the relationship between the relative use of conservation tillage and the real price of crude oil over the period 1963 to 1997.

Finally, Chapter 10 begins by noting that the environmental consequences of conservation tillage practices are an important issue concerning the impact of agricultural production on the environment. While it is generally recognized that water runoff and soil erosion will decline as no till and mulch tillage systems are used more extensively on cropland, what will happen to pesticide and fertilizer use remains uncertain. To gain some insight into this, the conservation tillage adoption decision is modeled. Starting with the assumption that this decision is a two step procedure - the first is the decision whether or not to adopt a conservation tillage production system and the second is the decision on the extent to which conservation tillage should be used - appropriate models of the Cragg and Heckman (dominance) type are estimated. Based on farm-level data on corn production in the United States for 1987, the profile of a farm on which conservation tillage was adopted is that the cropland had above average slope and experienced above average rainfall, the farm was a cash grain enterprise, and it had an above average expenditure on pesticides and a below average expenditure on fuel and a below average expenditure on custom pesticide applications. Additionally, for a farm adopting a no till production practice, an above average expenditure was made on fertilizer.

CHAPTER 2

AGRICULTURAL CHEMICAL USE

INTRODUCTION

Agricultural chemical use in the United States is comprised overwhelmingly of fertilizer-- more than 20 million tons have been applied annually in recent years--yet attention often focuses on the half-million ton of pesticides used each year. It is important to understand the nature and extent of agricultural chemical use, given the health, regulatory, and economic concerns that surround the issue. Alarms are expressed over acute and chronic poisonings by pesticides, detection of pesticide and fertilizer residues in food and water, or in the environment far from where they are applied. Others counter that detection alone do not imply serious risk. "In the face of extensive data gaps affecting risk assessment of pesticide residues, different stakeholders can plausibly argue that 'America's food supply is the safest in the world' or that it imposes 'intolerable risks' on children" (Pease et al. [1996]). Claims and counterclaims continue.

Uncertainties about the health effects of agricultural chemicals play an important role in regulatory concerns, such as those of recently enacted food quality and safe drinking water legislation, which have a bearing on agricultural chemical use. For example, under the Food Quality Protection Act (FQPA) of 1996, greater consideration of the effects of agricultural chemicals on infants and children, who are generally more sensitive to chemical exposures, must be given when setting Federal standards. Required under the FQPA and authorized under the Safe Drinking Water Amendments Act of 1996, greater attention must be given to estrogenic substances (those that may disrupt the human endocrine, or hormone, system). Some agricultural chemicals fall into that category. Both acts also include provisions to inform consumers about pesticides in foods and drinking water. Information on food residues is to be distributed to large retail grocers for public display, while information on residues detected in drinking water will be mailed or otherwise distributed to customers by their local water utility. Water quality regulations also pose an indirect concern about agricultural chemical use. As public water suppliers are required to meet Federal standards for agricultural chemicals in drinking water, local utilities (and their customers) must bear the cost for treating water sources that are contaminated by farming practices upstream.

Changes in agricultural chemical use, or pest resistance to pesticides, carry potentially significant economic impacts for input suppliers, producers, and consumers. This is because chemical use appears to be cost-effective for many producers and is widespread in the agricultural sector (e.g., nearly 100 percent of some major crops receive herbicide applications). Agricultural chemical use in 1995 accounted for about $17 billion in input costs, or more than 10 percent of total operating expenses (*Agricultural Outlook* [1996]). Agricultural pesticide use accounted for more than 70 percent of U.S. pesticide expenditures (Aspelin [1994]). (Similar data are unavailable for fertilizer use.) Further, all consumers are affected by farmers' production practices, if only by changing food costs. Even though farm prices account for a small share of retail food prices, small changes across large segments of the population can accrue to significant amounts.

Studies of banning, or pest resistance to, specific pesticides, classes of pesticides, or all pesticides show a range of economic losses to producers and consumers. These studies imply large benefits to pest control, a point on which there is little disagreement. A question for some is whether alternative means can be used to achieve better pest and nutrient management. To this end, the Administration has declared a goal of 75 percent adoption of Integrated Pest Management practices on U.S. crop acreage by the year 2000 to help reduce risks associated with pesticide use.

BACKGROUND

The growth of agricultural chemical use is an integral part of the technological revolution in agriculture that has generated major changes in production techniques, shifts in input use, and growth in output and productivity. The mechanization revolution of the 1930s and 1940s has been augmented since

1945 by a biological revolution in terms of fertilizer and pesticides (Carlson and Castle [1972]). That biological revolution continues.

Figure 2.1. Agricultural Input Use

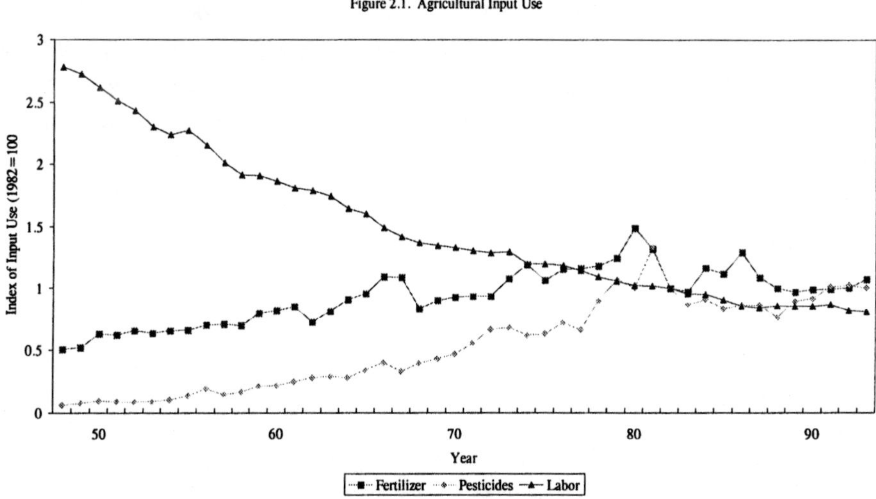

A consequence of these changes is that farmers use more agricultural chemicals but less labor. Indexes of total input use show that labor use (including hired, owner-operator, and unpaid family labor) fell at a 2.8 annual percentage rate between 1948 and 1993 (Figure 2.1).[1] Fertilizer use, on the other hand, increased by 1.6 percent annually while pesticide use grew by 6.1 percent annually.[2] It is evident from Figure 2.1, however, that fertilizer and pesticide use did not increase at a constant rate over the period 1948-1993. Between 1948 and 1981, for example, fertilizer use increased at an annual rate of 2.6 percent and was approximately unchanged between 1982 and 1993. Pesticide use increased annually at an 8.5 percent rate between 1948 and 1981 and subsequently have exhibited no identifiable trend. Agricultural productiv-

[1] This index of sectoral labor input accounts for changes in education and the composition of the labor force by incorporating the characteristics of individual workers.

[2] As will be discussed below, since the early 1960s the nitrogen content of fertilizer and the herbicide share of total pesticides has more than doubled. To account for such changes, quality differences measured in terms of common attributes have been incorporated into the indexes. For fertilizer, these attributes are nutrient content of the fertilizer materials. For pesticides, differentiation is based on physical characteristics such as toxicity, persistence in the environment, and leaching potential.

ity (total output divided by total factor inputs) expanded by 1.8 percent annually over this period (Ball and Nehring [1996], Ball et al. [1997])).

An alternative way of looking at changes in agricultural chemical use is to examine output per unit of input. From this perspective and based on indexes of aggregate output and total input use, output per unit of labor input increased at an approximately constant rate of 4.6 percent annually between 1948 and 1993 (Figure 2.2). In contrast, output per unit of fertilizer input decreased at annual rate of 0.7 percent between 1948 and 1981 and increased at a rate of 1.8 percent between 1982 and 1993 while output per unit of pesticide input decreased at a rate of 6.6 percent between 1948 and 1981 and increased slightly at a rate 0.6 percent annually between 1982 and 1993.

Figure 2.2. Output/Fertilizer Use, Output/Pesticide Use and Output/Labor Use

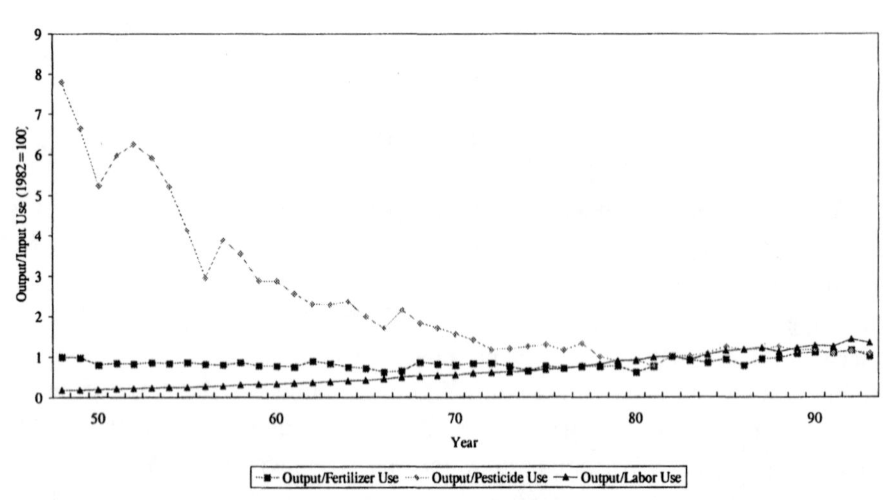

While the trends in agricultural chemical use are inherently interesting, a more relevant issue is what underlies the trends. A number of factors have contributed to the changes in the use trends of fertilizer and pesticides. In what follows, the elements that have precipitated these trends and the changes in the trends will be examined.

FERTILIZER USE TRENDS

From the European settlement of the United States until the 19th century, increased agricultural production came almost entirely from expanding the cropland base and mining the nutrients in the soil. With increased demand for agricultural commodities associated with the increase in the population of the United States, soil nutrient replacement to maintain or expand crop yields were desirable. Manure and other farm refuse were applied to the soils. Later, applications of manure were supplemented with fish, seaweed, peatmoss, leaves, straw, leached ashes, bonemeal, and Peruvian guano, materials that contained a higher percentage of nitrogen, phosphate, and potash than did manure (Wines [1985]). Crop rotations were also used to supply needed nutrients. Early crop rotations included combinations of row crops and small grains with hay and pasture over a 3 to 7 year cycle. Legumes in rotations fix nitrogen from the atmosphere into the soil, and rotations also helped improve soil moisture, control pests and increase yields (National Research Council [1989]). As manufacturing processes developed, production of synthetic chemical fertilizers like superphosphates and later urea and anhydrous ammonia replaced most natural fertilizers produced from recycled wastes. With the introduction of high analysis fertilizers like urea, anhydrous ammonia, and diammonium phosphate, monoculture or continuous same crop plantings for two or more years became popular with crops like corn, wheat, and soybeans.

Commercial fertilizers provide relatively low-cost nutrients needed to fully realize the yield potential of new high-yielding varieties (HYV) of crops (Ibach and Moyle [1971]). Since the early 1960s in response to the adoption of the HYV that are nitrogen intensive, U.S. yields per unit of land area for major crops have increased dramatically.[3] Corn yields, for example, increased from 62 bushels per acre in 1964 to 139 bushels in 1994 while wheat yields increased from 26 to 38 bushels during this same time period (U.S. Department of Agriculture [1994]). In 1964, the average application rate per fertilized corn acre was 58 pounds of nitrogen, 41 pounds of phosphate, and 35 pounds of potash. By 1994 this rate had increased to 129 pounds of nitrogen, 57 pounds of phosphate, and 80 pounds of potash.

[3] Nitrogen is a component part of amino acids, and hence, it is at the heart of protein molecules. Nitrogen fertilizer, in addition to increasing yields, also improves crop quality by increasing protein content. Crops such as corn, forage grasses, sorghum, and the small grains are well known for this characteristic.

Table 2.1. U.S. Commercial Fertilizer Use, 1960-95[1]

Year ending June 30[2]	Total fertilizer materials	Primary nutrient use			
		Nitrogen	Phosphate	Potash	Total[3]
	Million tons				
1960	24.9	2.7	2.6	2.2	7.5
1961	25.6	3.0	2.6	2.2	7.8
1962	26.6	3.4	2.8	2.3	8.4
1963	28.8	3.9	3.1	2.5	9.5
1964	30.7	4.4	3.4	2.7	10.5
1965	31.8	4.6	3.5	2.8	10.9
1966	34.5	5.3	3.9	3.2	12.4
1967	37.1	6.0	4.3	3.6	14.0
1968	38.7	6.8	4.4	3.8	15.0
1969	38.9	6.9	4.7	3.9	15.5
1970	39.6	7.5	4.6	4.0	16.1
1971	41.1	8.1	4.8	4.2	17.2
1972	41.2	8.0	4.9	4.3	17.2
1973	43.3	8.3	5.1	4.6	18.0
1974	47.1	9.2	5.1	5.1	19.3
1975	42.5	8.6	4.5	4.4	17.6
1976	49.2	10.4	5.2	5.2	20.8
1977	51.6	10.6	5.6	5.8	22.1
1978	47.5	10.0	5.1	5.5	20.6
1979	51.5	10.7	5.6	6.2	22.6
1980	52.8	11.4	5.4	6.2	23.1
1981	54.0	11.9	5.4	6.3	23.7
1982	48.7	11.0	4.8	5.6	21.4
1983	41.8	9.1	4.1	4.8	18.1
1984	50.1	11.1	4.9	5.8	21.8
1985	49.1	11.5	4.7	5.6	21.7
1986	44.1	10.4	4.2	5.1	19.7
1987	43.0	10.2	4.0	4.8	19.1
1988	44.5	10.5	4.1	5.0	19.6
1989	44.9	10.6	4.1	4.8	19.6
1990	47.7	11.1	4.3	5.2	20.6
1991	47.3	11.3	4.2	5.0	20.5
1992	48.8	11.4	4.2	5.0	20.7
1993	49.2	11.4	4.4	5.1	20.9
1994	52.3	12.6	4.5	5.3	22.4
1995	50.7	11.7	4.4	5.1	21.3

[1] Includes Puerto Rico.

[2] Fertilizer use estimates for 1960-84 are based on U.S. Department of Agriculture [1994 and earlier]. Data for 1985-95 are Tennessee Valley Authority [1995 and earlier] estimates.

Because the demand for fertilizer is a derived demand, fertilizer use is impacted by normal economic influences. A number of economic factors in both the output markets and input markets are responsible for the rise in the U.S. consumption of nitrogen, phosphate, and potash from 7.5 million nutrient tons in 1960 to a record high of 23.7 million nutrient tons by 1981, an increase of over 217 percent (Table 2.1). In the output markets during the 1970s and early 1980s, commodity prices were above their historical average due to Government price support programs. Corn, for example, averaged $2.10 per bushel between 1970 and 1979, much greater than the average of $1.17 per bushel a decade earlier.[4] Thus, with farmers equating the value of the marginal product of fertilizer to the price of fertilizer[5] and with the value of the marginal product being elevated, fertilizer use expanded.

On the input side, changing relative factor prices led to the substitution of fertilizer for other factors because it became relatively less expensive. With the exception of a few years (e.g., 1974 and 1975), the relative change in the price of fertilizer is consistently smaller than the change in the price of other inputs leading, to the extent possible, substitution of fertilizer for other factors of production.

Another strong impetus for fertilizer substitution in the decade of the 1970s was the increase in the price of farmland. Between 1970 and the peak year of 1982, farmland values increased from a nominal amount of $157 to $715, an aggregate increase of 355 percent or an annual increase of 11.1 percent. Over the same time period, the price of fertilizer increased by slightly more than 195 percent or an annual increase of 5.7 percent. As a consequence, farming during this period was on a more intensive margin relying on increased use of fertilizer as reflected in Figure 2.1.

Next, U.S. consumption of nitrogen, phosphate, and potash is tied to total planted crop acreage. As crop acreage increases, fertilizer consumption increases as well. Consequently, Government programs that encourage farmers to divert acreage from production can significantly affect fertilizer consumption. Fertilizer use, however, also changes as application rates on specific crops and the proportion of the acreage of those crops fertilized change over

[4] These are nominal prices. The standard deviations are 0.07 for 1960-1969 and 0.58 for 1970-1979.

[5] This is a simple result from neoclassical microeconomic theory.

time. Thus, for example and as alluded to previously, the increase in corn production has been a major contributor to the growth in the use of plant nutrients. In 1964 corn accounted for 37 percent of all nitrogen fertilizer used and 34 percent of all plant nutrients (including nitrogen fertilizer) used while accounting for 18 percent of planted acreage. By 1997 it was responsible for 50 percent of all nitrogen and 49 percent of all plant nutrients used while accounting for approximately 24 percent of total planted acreage.[6] The increase in fertilizer use on corn especially through 1985 was due in part to greater application rates but it was also enhanced by a rise in the proportion of corn acres fertilized with nitrogen and an increase in planted acreage. Application rates for nitrogen and potash were 140 percent greater in 1985 (the peak year for nitrogen fertilizer application until recently) than in 1964, while the rate of phosphate applied rose by 46 percent. During this period, corn planted acreage increased by about 27 percent.

Total nutrient use has fallen somewhat from its peak in 1981 along with total crop acreage, particularly in 1983 as a result of the PIK programs. It totaled 22.4 million nutrient tons in 1994. Land values have also dropped from their peak in 1982 ranging from around $550 to $600 per acre in the early 1990s. Commodity prices have remained relatively stable during this time due to Government price support programs. For example, nominal corn prices averaged $2.40 per bushel between 1980 and 1994 with a standard deviation of 0.40.[7]

Nitrogen, phosphate, and potash all shared in the dramatic increase in nutrient use between 1960 and 1981. The relative use of nitrogen, however, increased much more rapidly. Nitrogen consumption stood at 2.7 million nutrient tons in 1960, or 36.7 percent of total nutrient consumption. By 1981, nitrogen use had increased to 11.9 million nutrient tons. Nitrogen application rates on corn increased from 58 pounds per acre in 1964 to 137 pounds in 1981 and the percent of corn acres receiving nitrogen increased from 85 to 97. Nitrogen use stood at 12.6 million tons in 1994, its highest level on record, and accounted for 56 percent of total nutrient consumption.

While nitrogen's share of total plant nutrient consumption increased over this period, phosphate's share declined from 34.5 percent in 1960 to 20.2 percent by 1994 while per acre application rates on crops like corn increased.

---

[6] In 1964 total cropland acreage was 292 million acres. In 1997, total acreage was 321 million acres.

[7] In this instance, the standard deviation can be used as an indicator of price variability.

Phosphate use, which increased from 2.6 million nutrient tons in 1960 to a peak of 5.6 million nutrient tons in 1977, has basically followed a downward trend since 1979. One primary reason for this is that plant needs can be met either by direct application of commercial fertilizer or by allowing the crop to utilize residual nutrients already in the soil. Since the late 1970s, U.S. farmers have been taking advantage of residual soil supplies which had been built up on some fields by previous fertilizer and manure applications thereby reducing their purchases of phosphate (Wallingford [1992]).

Potash consumption, historically below that of both nitrogen and phosphate, surpassed phosphate consumption for the first time in 1977 and will likely hold this position. Total use of potash increased from 2.2 million nutrient tons in 1960 to 6.3 million nutrient tons in 1981. Since then the consumption of potash has fallen, paralleling phosphate consumption. While the shares of total primary nutrient consumption held by nitrogen and phosphate have increased and decreased, respectively, potash's share has remained stable.

An adequate supply of nutrients and maintenance of proper soil acidity are essential to crop growth. Ideally, soil nutrients should be available in the proper amounts at the time the plant can use them. Relatively heavy use of nitrogen and some other fertilizers, however, can lead to soil acidification, other changes in soil properties, and offsite environmental problems. Fertilizers are sometimes over-applied. When this occurs, the total amount of plant nutrients available to growing crops not only exceeds the need or ability of the plant to absorb them but exceeds the economic optimum as well. Estimates of crop absorption of applied nitrogen range from 25 to 70 percent and generally vary as a function of plant growth and health and the method and timing of nitrogen application (National Research Council [1989]). Crops are much more likely to more fully utilize nitrogen that is properly timed. Unused nitrogen can be immobilized, denitrified, washed into surface water, or leached into groundwater (Huang et al. [1992]).

Focusing on the nitrogen mass balance through proper timing avoids supplying an excess that cannot be used by plants and many become a potential source of environmental contamination. Since the late 1980s, alternative production practices tied to nutrient mass balance have been emphasized. Best management practices (BMPs) have been introduced which involve all aspects of crop production, including cultural practices such as cultivation, fertilization use and timing, split fertilizer applications, nutrient credits for crop residues and use of manure, precision farming, postharvest management of fields, scouting for pests, tillage practices, the use of genetically improved pest-resistance varieties, crop rotations, and the use of biological controls. The concepts of BMPs is that of an economic threshold below which pest populations or damage is tolerated. Included in this are the value of legumes in fixing nitrogen, improving soil quality, reducing erosion, and increasing yields of subsequent crops. Total fertilizer use and application rates on major crops during the late 1980s and early 1990s have reflected these practices. While their precise impact on fertilizer use is difficult to quantify, the widespread adoption of such cropping practices have served to improve fertilizer use efficiency. This has impacted the fertilizer use trend, in some cases increasing use and in other cases reducing use.

PESTICIDE USE TRENDS

Prior to the development of synthetic pesticides following World War II, a farmer's solution to weed, insect, and disease problems was limited to physical and cultural practices which often failed and resulted in crop damages and yield reductions. Weed control was accomplished with tillage implements, mowing, site selection, use of seeds free of weedseeds, and often even with the use of hands and hand tools. While attempts were made to control insect pests and diseases through seed selection, crop rotations, trap crops, adjustment of planting dates, or other cultural practices, severe yield losses and abandonment of production was still common when heavy pest infestations occurred.

In the thirty-five years following World War II, chemical pest control was widely adopted on most crops. During this time, pesticides were applied to an increasing share of the acres and at increasing application rates per acre. Public and private research efforts after World War II introduced new pesticides along with many other complementary innovations to increase yields and reduce farm labor needs. The availability of 2,4-D and DDT in the 1940s and the expectation for more chemical controls offered the potential to replace many labor intensive cultural practices with a herbicide. Genetic improvements in seeds that were resistant to both pests and pesticides and improved tillage equipment that allowed more timely and effective tillage also became available to reduce pest losses and labor needs. Rising land values from a growing demand for agricultural products and higher industrial wage rates luring workers out of agriculture provided an economic stimulus for farmers to adopt these and many other labor-saving and yield- increasing technologies over the next thirty years. Rising energy prices in the 1970s and the additional policy emphasis for soil conservation provided added impetus to adopt herbicides as a substitute for high energy-using tillage systems and frequent cultivations.

A turning point in pesticide use came in the early 1980s when high interest rates, falling land values, large land set-aside programs, and low farm incomes resulted in fewer planted acres and reductions in purchased inputs. Although little or no trend is associated with the total quantity of pesticides used since the early 1980s significant changes have occurred in the type and intensity which are masked by the aggregate numbers. Herbicide usage on most crops continue to edge toward 100 percent of the planted acreage. Average application rates declined for several major crops, but the number of herbicide treatments and number of herbicide ingredients applied to each acre increased. Larger quantities of fungicides and other pesticides were used resulting from both increases in treated acres and more intense treatments. Some of these underlying trends are addressed in this report.

Studies of pesticide productivity report a wide range of economic returns for each dollar spent on pesticides, but most studies report that the marginal return of using pesticides is much greater than the marginal costs. Because pesticide use often varies simultaneously with the adoption of many other inputs and practices, such as fertilizer and improved seeds and mechanization, the independent contribution of pesticides is difficult to accurately estimate. While most studies report positive marginal returns in the range of $1 to $3 for each dollar spent on pesticides, some are much higher while others have shown the marginal cost exceeding the return (Headley [1968], Padgitt [1969], Hawkins et al. [1977], Carlson [1977], Duffy and Hanthorn [1978]). More recent studies, however, tend to show lower levels of returns than studies conducted in the 1960s. This decline in the marginal productivity as pesticide use increased across most agricultural crops is expected.

Although the availability of synthetic pesticides greatly increased a producer's ability to control pests, reduce labor, expand crop acreage, and increase productivity, it was not a solution without problems. Growing concerns about agriculture's continued reliance on chemical pest control methods raised issues about health risks from residues on food or pesticides in drinking water. Environmentalists have concerns about pesticide impacts on wildlife, sensitive eco-systems, and changes in the biodiversity of plant and animal life. Farmers have their own concerns about their health from pesticide exposure during mixing and application and the development of pesticide resistant species from continuous use of the same ingredients. These concerns of how pesticide toxicity and their potential to impact on the environment has changed over time may be of equal or more interest than how they contribute to agricultural productivity.

As illustrated in Figure 2.1, pesticide use in agriculture increased until the early 1980s, coinciding with the growth in crop acres and a larger share of crop acres receiving pesticide treatments. Since 1982, following the peak in crop acres planted, the quantity of pesticide active ingredients used by farmers has fluctuated between 800 and 860 million pounds. The annual fluctuations have partially been in response to changes in crop acres, government set-aside requirements, and pest infestations, but also reflects some trends in the intensity of use per acre and the affect of newer products applied at different rates than the older products they replaced.

Pesticide products are classified as herbicides, insecticides, fungicides, and "other" pesticides for reporting purposes of this chapter. Detail about the type and quantity of pesticides applied to major field crops, fruits and vegetables is provided in Table 2.2.

While all of the conventional measurement units provide a relative indication of intensity of use, they fail to account for differences in product characteristics or changes in the mix of products over time. Because each pesticide ingredient can differ greatly in its toxicity, persistence, and mobility, other measurement units are needed to reflect any changes in the health or environmental risks posed by pesticides. Reported later in this chapter are constructed measures which adjusts for different toxicity characteristics and illustrates how these characteristics have changed over time. Due to canceled pesticides, changes in use restriction, and shifts to less toxic chemicals, the perception that health and environmental risks parallel the aggregate changes in pesticide quantities is not necessarily valid.

Table 2.2. Estimated quantities of pesticide active ingredients applied to selected U.S. crops, 1964-95[1]

Commodities	1964	1966	1971	1976	1982	1990	1991	1992	1993	1994	1995
1,000 pounds of herbicides											
Corn	25,476	45,970	101,060	207,061	243,409	217,500	210,200	224,363	201,997	215,636	186,314
Cotton	4,628	6,526	19,610	18,312	20,748	21,114	26,032	25,773	23,567	28,565	32,873
Wheat	9,178	8,247	11,622	21,879	19,524	16,641	13,561	17,387	18,304	20,708	20,054
Sorghum	1,966	4,031	11,538	15,719	15,738	13,485	14,156	na	na	na	na
Rice	2,559	2,819	7,985	8,507	14,089	16,139	16,092	17,665	na	na	na
Soybeans	4,208	10,409	36,519	81,063	133,240	74,400	69,931	67,358	64,092	69,257	68,126
Peanuts	2,894	2,899	4,374	3,366	4,927	4,070	4,510	na	na	na	na
Potatoes	1,297	2,220	2,178	1,764	1,636	2,361	2,547	2,152	2,504	2,866	2,894
Other Vegetable	2,194	3,488	3,361	5,419	4,345	4,916	4,712	5,850	5,743	6,138	6,122
Citrus	207	353	546	4,756	6,289	5,652	6,076	5,545	5,086	4,793	4,665
Apples	278	389	156	575	649	396	429	419	445	604	769
Other Fruit	692	1,782	615	0	504	1,659	1,690	1,687	1,774	1,877	1,891
1,000 pounds of insecticides											
Corn	15,668	23,629	25,531	31,979	30,102	23,200	23,036	20,866	18,479	17,349	14,956
Cotton	78,022	64,900	73,357	64,139	19,201	13,583	8,159	15,307	15,429	23,882	30,039
Wheat	891	876	1,712	7,236	2,853	970	208	1,153	152	2,031	910
Sorghum	788	767	5,729	4,604	2,559	1,085	1,140	na	na	na	na
Rice	284	312	946	508	565	161	309	178	na	na	na
Soybeans	4,997	3,217	5,621	7,866	11,621	0	445	359	346	203	515
Peanuts	5,518	5,529	5,993	2,439	1,035	1,726	1,913	na	na	na	na
Potatoes	1,456	2,972	2,770	3,261	3,776	3,591	3,597	3,514	3,943	4,459	3,109
Other vegetable	8,290	8,163	8,269	5,671	4,465	4,709	4,466	5,482	5,306	5,592	5,576
Citrus	1,425	2,858	3,049	4,604	5,306	2,811	3,977	4,538	5,271	5,110	5,143

Table 2.2. Estimated quantities of pesticide active ingredients applied to selected U.S. crops, 1964-95[1] (continued)

Apples	10,828	8,494	4,831	3,613	3,312	3,691	4,013	3,909	4,150	3,843	3,573
Other fruit	1,727	4,131	2,569	3,361	2,016	4,837	4,928	4,919	5,023	5,408	5,533
1,000 pounds of fungicides											
Corn	0	0	0	20	69	0	0	0	0	0	19
Cotton	171	376	220	49	200	988	701	785	684	1,065	1,045
Wheat	0	0	0	862	1,088	172	73	1,154	688	1,012	500
Sorghum	0	0	0	0	0	0	0	na	na	na	na
Rice	0	0	0	0	80	194	426	388	na	na	na
Soybeans	0	0	0	176	71	0	0	85	0	45	13
Peanuts	1,106	1,108	4,431	6,834	4,739	7,321	8,114	6,725	na	na	na
Potatoes	3,229	3,531	4,124	4,168	4,031	2,808	3,172	3,616	4,369	6,358	7,973
Vegetables	4,530	4,093	5,667	5,051	6,692	12,917	13,126	17,260	18,720	21,885	21,820
Citrus	4,929	4,056	9,257	5,897	4,881	2,555	3,598	3,429	3,322	3,582	4,019
Apples	7,750	8,496	7,207	6,489	5,667	4,177	4,544	4,377	4,599	4,623	4,692
Other fruit	1,558	2,685	2,833	3,921	2,520	4,146	4,224	4,216	4,206	4,477	4,535
1,000 pounds of other pesticides											
Corn	76	546	443	483	130	0	0	0	0	0	0
Cotton	12,431	14,207	18,696	12,682	9,347	15,188	15,457	15,781	12,658	15,616	19,733
Wheat	0	47	245	0	0	0	0	0	0	0	0
Sorghum	0	40	0	266	44	0	0	na	na	na	na
Rice	0	0	0	0	17	0	0	109	na	na	na
Soybeans	0	49	52	2,030	2,430	0	0	0	0	0	0
Peanuts	6,990	7,005	471	1,188	1,627	2,364	2,620	na	na	na	na
Potatoes	91	9	6,397	8,576	15,188	35,069	45,626	49,671	57,494	79,809	72,928
Other vegetable	5,819	569	3,435	5,061	6,206	17,283	17,998	24,189	27,524	33,408	33,307

Table 2.2. Estimated quantities of pesticide active ingredients applied to selected U.S. crops, 1964-95[1]
(continued)

Citrus	1,539	681	1,280	214	7	10	15	31	49	108	179
Apples	1,037	1,079	548	574	421	73	73	66	65	79	93
Other fruit	386	1,560	614	504	276	282	281	27	625	1,167	
1,000 pounds of all pesticide types											
Corn	41,220	70,145	127,034	239,543	273,710	240,700	233,235	245,229	220,476	232,985	201,289
Cotton	95,252	86,009	111,883	95,182	49,497	50,873	50,349	57,646	52,338	69,128	83,689
Wheat	10,069	9,170	13,579	29,977	23,465	17,782	13,842	19,694	19,144	23,751	21,464
Sorghum	2,754	4,838	17,267	20,589	18,341	14,570	15,296	na	na	na	na
Rice	2,843	3,131	8,931	9,015	14,751	16,494	16,827	18,340	na	na	na
Soybeans	9,205	13,675	42,192	91,135	147,362	74,400	70,376	67,802	64,438	69,505	68,655
Peanuts	16,509	16,541	15,268	13,827	12,327	15,482	17,157	na	na	na	na
Potatoes	6,073	8,732	15,470	17,769	24,631	43,830	54,942	58,953	68,309	93,492	86,904
Other vegetable	20,833	16,313	20,732	21,202	21,707	39,824	40,302	52,781	57,292	67,024	66,825
Citrus	8,100	7,948	14,132	15,471	16,483	11,028	13,666	13,544	13,729	13,594	14,006
Apples	19,893	18,458	12,742	11,251	10,049	8,337	9,059	8,771	9,260	9,149	9,128
Other fruit	4,364	10,158	6,631	7,283	5,544	10,919	11,123	11,103	11,030	12,386	13,126

[1] Estimates are constructed for the total U.S. acreage of the selected commodities. In years when the surveys did not include all states producing the crop, the estimates assume similar use rates for those states.
Source: Economic Research Service [1997].

Table 2.3. Pesticide treatments on major field crops in major production states, 1990-1995

Data Item	Units	1990	1991	1992	1993	1994	1995
Corn:							
Planted area	(1,000 ac.)	58,800	60,350	62,850	57,350	62,500	55,850
Land receiving herbicides	(Percent)	94.8	95.5	96.9	97.6	98.0	97.5
Ave. number of treatments	No.	1.4	1.4	1.4	1.4	1.5	1.5
Ave. number of ingredients	No.	2.2	2.1	2.3	2.3	2.5	2.4
Ave. number of acre-treatments	No.	2.2	2.2	2.3	2.4	2.5	2.5
Amount of herbicide applied	(lbs./ac.)	3.24	2.97	2.98	2.94	2.79	2.76
Before or at plant only	(percent)	39.3	38.4	33.0	34.7	29.4	30.4
After plant only	(percent)	29.1	34.1	36.4	36.8	38.1	37.9
Both	(percent)	26.4	23.0	27.2	25.6	30.2	29.0
Land receiving insecticides	(Percent)	32.3	30.4	28.7	28.2	26.5	26.0
Ave. number of treatments	No.	1.1	1.1	1.1	1.1	1.1	1.1
Ave. number of ingredients	No.	1.1	1.1	1.1	1.0	1.1	1.1
Ave. number of acre-treatments	No.	1.1	1.1	1.1	1.1	1.1	1.1
Amount of insecticide applied	(lbs./ac.)	1.18	1.04	0.95	0.90	0.83	0.75
Before or at plant only	(percent)	25.8	22.7	22.5	22.3	18.5	17.9
After plant only	(percent)	4.4	5.6	4.9	5.2	6.5	6.9
Both	(percent)	2.1	2.1	1.3	0.7	1.5	1.2
Soybeans:							
Planted area	(1,000 ac.)	39,500	42,050	41,350	42,500	43,750	45,150
Land receiving herbicides	(Percent)	95.8	96.8	98.2	97.5	98.4	97.7
Ave. number of treatments	No.	1.5	1.5	1.6	1.6	1.7	1.7
Ave. number of ingredients	No.	2.3	2.3	2.4	2.5	2.7	2.7
Ave. number of acre-treatments	No.	2.3	2.3	2.4	2.5	2.8	2.8
Amount of herbicide applied	(lbs./ac.)	1.39	1.27	1.14	1.11	1.14	1.09

Table 2.3. Pesticide treatments on major field crops in major production states, 1990-1995 (cont'd)

Before or at plant only	(percent)	44.2	39.1	35.9	27.7	28.1	23.4
After plant only	(percent)	20.1	26.1	27.9	29.7	28.5	32.1
Both	(percent)	31.5	31.6	34.4	35.1	41.6	42.2
Cotton:							
Planted area	(1,000 ac.)	9,730	10,860	10,200	10,360	10,023	11,650
Land receiving herbicides	(Percent)	94.7	92.0	90.8	92.0	93.5	97.7
Ave. number of treatments	No.	2.1	2.3	2.5	2.5	2.6	2.7
Ave. number of ingredients	No.	2.3	2.5	2.7	2.7	2.7	2.8
Ave. number of acre-treatments	No.	2.7	2.9	3.2	3.2	3.5	3.3
Amount of herbicide applied	(lbs./ac.)	1.79	2.01	2.11	2.01	2.23	2.03
Before or at plant only	(percent)	57.7	52.0	49.1	45.0	41.2	46.1
After plant only	(percent)	5.7	5.1	9.2	9.5	6.2	6.7
Both	(percent)	31.3	34.5	32.5	37.5	46.1	44.7
Land receiving insecticides	(Percent)	1/	66.3	64.9	64.8	71.0	76.2
Ave. number of treatments	No.	1/	3.1	4.5	4.9	5.7	6.2
Ave. number of ingredients	No.	1/	2.3	3.2	3.4	3.5	3.8
Ave. number of acre-treatments	No.	1/	3.7	6.0	6.6	7.7	7 .9
Amount of insecticide applied	(lbs./ac.)	1/	1.13	1.83	2.06	2.48	2.36
Land receiving other pest.	(Percent)	1/	56.5	47.3	63.5	66.8	57.0
Ave. number of treatments	No.	1/	1.8	1.6	1.6	1.7	2.1
Ave. number of ingredients	No.	1/	2.0	2.0	1.9	2.0	2.1
Ave number of acre-treatments	No.	1/	2.4	2.3	2.2	2.7	2.7
Amount of other pesticides applied	(lbs./ac.)	1/	1.63	2.34	1.79	1.72	2.40

Table 2.3. Pesticide treatments on major field crops in major production states, 1990-1995 (cont'd)

Wheat:							
Planted area	(1,000 ac.)	57,800	47,500	53,540	54,400	52,630	51,370
Land receiving herbicides	(Percent)	52.9	49.8	52.8	60.1	64.9	69.3
Ave. number of treatments	No.	1.2	1.2	1.2	1.2	1.2	1.2
Ave. number of ingredients	No.	1.7	1.9	1.9	2.0	2.0	2.1
Ave. number of acre-treatments	No.	1.7	1.9	1.9	2.1	2.1	2.1
Amount of herbicide applied	(lbs./ac.)	0.43	0.43	0.42	0.41	0.44	.40
Before or at plant only	(percent)	2.7	2.6	3.7	3.5	4.5	3.1
After plant only	(percent)	45.9	42.4	45.9	52.2	54.4	61.5
Both	(percent)	4.4	4.8	3.1	4.2	4.6	4.4
Land receiving insecticides	(Percent)	4.8	6.6	3.7	1.6	7.4	3.9
Ave. number of treatments	No.	1.1	1.0	1.0	1.0	1.1	1.0
Ave. number of ingredients	No.	1.1	1.1	1.1	1.0	1.0	1.0
Ave. No. of acre-treatments	No.	1.1	1.1	1.1	1.0	1.1	1.1
Amount of insecticide applied	(lbs./ac.)	0.49	0.28	0.39	0.25	0.41	0.36
Before or at planting only	percent)	0.1	0.3	.	0.1	0.2	0.0
After planting only	(percent)	4.7	6.3	3.7	1.5	7.2	3.7
Fall Potatoes:							
Planted area	(1,000 ac.)	1,087	1,123	1,064	1,114	1,140	1,147
Land receiving herbicides	(Percent)	81.4	80.8	81.7	82.2	83.5	86.4
Ave. number of treatments	No.	1.3	1.4	1.3	1.3	1.4	1.4
Ave. number of ingredients	No.	1.6	1.7	1.7	1.7	1.8	1.9
Ave. number of acre-treatments	No.	1.7	1.8	1.7	1.8	1.9	2.0
Amount of herbicide applied	(lbs./ac.)	2.15	2.29	1.94	2.06	2.42	2.40
Before or at plant only	(percent)	15.8	12.5	13.9	14.1	16.0	9.6

Table 2.3. Pesticide treatments on major field crops in major production states, 1990-1995 (cont'd)

After plant only	(percent)	59.7	61.1	62.6	61.5	58.0	72.2
Both	(percent)	5.7	7.2	5.2	6.6	9.5	4.7
Land receiving insecticides	(Percent)	88.5	91.8	90.0	88.4	87.7	87.8
Ave. number of treatments	No.	2.0	2.2	2.3	2.2	2.7	2.5
Ave. number of ingredients	No.	1.8	1.9	2.0	2.0	2.1	1.9
Ave. number of acre-treatments	No.	2.1	2.4	2.6	2.5	3.1	2.6
Amount of insecticide applied	(lbs./ac.)	3.15	2.81	2.89	2.90	3.49	2.55
Before or at plant only	(percent)	17.6	13.0	13.6	13.8	15.9	15.8
After plant only	(percent)	51.9	57.5	59.5	58.5	59.0	53.2
Both	(percent)	19.0	21.3	16.9	16.1	12.8	18.8
Land receiving fungicides	(Percent)	69.0	68.6	72.1	76.3	79.8	84.8
Ave. number of treatments	No.	2.7	2.7	3.1	3.4	4.2	6.1
Ave. number of ingredients	No.	1.4	1.5	1.9	2.1	2.3	2.7
Ave. number of acre-treatments	No.	3.2	3.6	4.2	4.7	6.1	7.5
Amount of fungicide applied	(lbs./ac.)	3.17	3.42	3.93	4.22	5.61	6.75
Land receiving other pest.	(Percent)	34.6	44.9	43.1	52.9	59.9	57.1
Ave. number of treatments	No.	1.3	1.3	1.4	1.3	1.4	1.6
Ave. number of ingredients	No.	1.1	1.2	1.3	1.2	1.2	1.3
Ave. number of acre-treatments	No.	1.4	1.3	1.5	1.4	1.5	1.6
Amount of other pesticides applied	(lbs./ac.)	73.38	71.24	84.43	74.56	94.36	92.74

1/ The 1990 survey for cotton only collected herbicide treatments.
Source: Economic Research Service [1997].

(A) HERBICIDES

Effective weed control usually results from a combination of cultural, tillage, and chemical practices. Most producers no longer rely on cultural and tillage implements alone. Alternate crops in rotation, which offer some weed control benefits, often do not provide high profits. Conventional tillage systems have high cost for implements, fuel, and labor, and are sometimes limited when they can be can be successfully applied due to adverse weather. Herbicides in combination with other practices are often more cost effective, provide improved weed control over a longer period, reduce the risk of poor weed, and result in higher profits. Research and experience with different herbicide products and applications rates allow producers to specifically tailor herbicide treatments to fit their unique situations. Among the factors to be considered in selecting herbicide ingredients, formulations, application methods and timing include soil texture, soil pH, organic content, soil moisture at application, soil incorporation requirements, weed species and pesticide resistant weeds, crop variety tolerance, following crop, and potential drift onto other fields.

Herbicides continue to dominate as the leading type of pesticide, accounting for 61 percent of all pesticides in 1994 (Table 2.2). The widespread adoption and use of herbicides, especially on corn and soybeans, accounted for much of the increase in herbicide use during the 1960s and 1970s. Total herbicide use increased from about 55 million pounds in 1964 to 465 million pounds in 1982. With this growth, herbicides replaced insecticides as the leading pesticide type. During this period the percent of corn and soybean acres receiving herbicides grew from about 20 percent to nearly 90%. Since the mid-1970s, herbicide usage on all the surveyed crops, except wheat, has edged toward 100 percent.

While there has been only small changes in pounds of herbicide use in recent years, there are some underlying trends in the share of acres treated, application rates, and the timing of applications as well as changes in applied active ingredients (Table 2.3). For example, winter wheat acres treated with herbicides increased from 33 percent in 1990 to 54 percent in 1995. Herbicide application rates declined 22 percent on soybean and 15 percent on corn, but increased 13 percent on cotton. While application rates on corn and soybeans dropped, the number of herbicide treatments and number of different ingredients applied on each treated acre increased as producers abandoned some ingredients applied at higher rates and adopted others applied at lower rats. The number of herbicide acre-treatments, which accounts for both the number of treatments and ingredients applied, is up for all major field crops. Combinations of herbicide ingredients along with more frequent treatments can provide a broader spectrum of weed control, reduce the potential for development of pesticide resistant weeds, or reduce carry-over affects to following crops.

Herbicide treatments after planting have increased on corn, soybean, and cotton and should provide some environmental quality benefits. For weed infestations which occur only with specific weather or in cycles, intervention treatments after the crop and weeds have emerged and reached threshold levels are often advocated as a pesticide reduction strategy. Also, post emergence herbicides usually have little residual soil activity and are less likely to have extended environmental impacts. The 1995 corn and soybean acreage which was treated only after planting is up about 10 percentage points from five years earlier. Except for the increased use of "burndown" herbicides associated with no-till systems, other preplant herbicides, which have greater soil residual activity and environmental risk, declined in use. These before or at planting applications are usually preventive treatments based on previous weed problems which are likely to recur. Some producers also find preventive treatments desirable because the appropriate timing of intervention treatments is not always possible because of weather or other factors.

(B) INSECTICIDES

Damaging insect populations can vary annually depending on weather, pest cycles, cultural practices such as crop rotations and destruction of previous crop residues, and other factors. Insecticide use includes both preventative treatments which are applied before infestation levels are known, and intervention treatments which are based on monitored infestation levels and expected crop losses. During the 1950s insecticide use became well established in the production of many crops susceptible to insect damages, including cotton, tobacco, and many fruits and vegetables and accounted for a dominate share of all pesticides. The total quantity of insecticides applied in the mid- to late 1970s significantly exceeded current use (Table 2.2). Insecticides used on major field crops, citrus and apples now account for about 10% of all pesticide materials. The estimated use of 62.6 million pounds in 1994 was less than half the quantity used in 1976 on these crops. The drop from earlier years is primarily due to the replacement of heavier organochlorine insecticides, such as DDT, aldrin, chlordane, and toxaphene, with synthetic pyrethroids and other insecticides that are commonly applied at much lower rates than those used in earlier years. The organochlorine insecticides were banned in the 1960s and 1970s for most agricultural uses due to their long run persistence in the environment and the health and environmental risk.

Insecticides used on corn and cotton now account for about two-thirds of all insecticides applied to the selected crops reported in Table 2.2. Total amounts applied to corn have decreased in recent years while that applied to cotton has increased. About one-fourth of the corn acres is currently treated with a single acre-treatment of insecticide but the share of acres treated has declined in recent years (Table 2.3). The primary use of insecticides on corn is for corn rootworm and other soil insects on continuous corn. These applications are most often preventive treatments applied at planting. Economic threshold for intervention treatments later in the growing season have not been developed for many other corn insect pests. Relatively few corn acres, however, receive insecticide treatments later in the growing season.

Most cotton, except for regions in Texas, is treated with insecticides. Only about 50 to 60 percent of the cotton grown in Texas receives insecticide, but in other major cotton production states, most of the acres receive multiple treatments during the growing season. The intensity of treatments are less in states with successful boll weevil eradication programs, however these areas still have other insect infestations requiring treatments. Beside boll weevil, cotton boll worms, tobacco budworms, and thrips are common target pest for insecticide treatments. Since 1991, not only has the share of cotton acres treated increased, but the intensity of the treatments has also increased. The total insecticide quantity applied per acre as well as the average number of acre-treatments doubled in the four years between 1991 and 1995. Unlike corn, economic thresholds have been developed for many cotton insects. Nearly all cotton acres are scouted for pests, except for regions in Texas where insect problems generally do not require insecticide treatments, and the majority of producer indicate that economic thresholds are used to make insecticide intervention decisions.

Insecticides applied on peanuts, fruit, potatoes and other vegetables account for most of the remaining one-third of the insecticide quantities reported in Table 2.2. Damaging insect infestation is normally low and requires little or no treatment on soybeans, sorghum, and rice. Control of Russian wheat aphid in Colorado, Oklahoma, and Texas is the most frequent use of insecticide on wheat.

(C) Fungicides and Other Pesticides

Fungicides represent the smallest class of pesticides used in agriculture. They only account for about 4% of the quantity of all pesticide types but are critical for the production of many high-valued fruit and vegetable crops. They are mostly used to control diseases which affect the health of the plant, but are also applied to prevent diseases affecting the appearance and quality of the fruit. The effectiveness of fungicides is short-lived when exposed to weather, consequently they often have to be re-applied several times during the growing season which partially accounts for their high seasonal application rates. Except for potatoes and other vegetables, the quantities applied to most crops have not changed significantly. Fungicide use on potatoes and other vegetables approximately doubled in the last four years with increased incidence of disease. For potatoes, the number of acres treated with fungicides is up an additional 15 percent during the last four years and the average number of application during the growing season more than doubled.

Pesticides classed as "other pesticides" represent a wide range of uses and diverse kinds of ingredients and have experienced the largest growth of all pesticide types in recent years. There use has increased about 22 percent each year since 1990 and they now account for about 20 percent of pesticides used on the crops reported in Table 2.2. These other pesticide products include soil fumigants, defoliants, and chemicals used to obtain special effects, such better coloration, shorter and denser plant structures, or controlled time of maturity. These products often aid the mechanical harvesting of crops through defoliation, simultaneous fruit setting and ripening, and the development of a more homogeneous plant structure.

Because of the large diversity of products and wide range in application rates for ingredients in the "other pesticide" class, changes in total quantities can easily misrepresent the overall trends. Included in this class are chemicals such as metam sodium, sulfuric acid, and dichloropropene, which are normally applied at several hundred pounds per acre. Small acreage changes in the use of such commodities can grossly distort the significance of the overall trends in the use of these pesticide types. Although the total pounds of other pesticides is up nearly 90% between 1990 and 1994 (Table 2.2), much of the increase stems from a small increase in potato acreage treated with soil fumigants and desiccants.

TRENDS IN PESTICIDE INGREDIENT USE

Many pesticides registered 25 or more years ago continue to have widespread usage. Atrazine, 2,4-D, and dicamba are among those products registered over 25 years ago and in 1995 were still the three most used herbicides in number of treated acres. Atrazine, registered in 1958, is mostly used on corn and sorghum to control broadleaf and several grass weeds. Atrazine, which is soluble in water and does not readily metabolize, has been found in wells and surface water sources in several corn production areas and is among the triazine herbicides under special review by the EPA because of its potential to cause cancer. Atrazine was reported used on 67 percent of the corn acres in 1976 and the share of corn acres treated has remained close to that level with 65 percent of the corn reported treated in 1995. However, over this period application rates have dropped with the growing concerns over residues leaching to groundwater and development of resistant weed species. Since 1976 average application rates dropped from 1.5 pounds per acre to 1.2 pounds in 1990, and to 1.1 pounds in 1995.

The herbicide 2,4-D was registered in 1948 and continues to be used as a post emergence treatment for many perennial and biennial broadleafs weeds on corn, sorghum, wheat, fruit orchards, and several other commodities. Although 2,4-D usage on major field crops is down from the 1970s, it is still the most commonly applied ingredient on wheat and is applied on 12 to 14 percent of the corn acres. It is often used in combination with other herbicides ingredients to provide a broader spectrum of weed control.

Dicamba is another commonly used broadleaf herbicide registered over 30 years ago and continues in use on corn and wheat. Although the total quantity of dicamba applied is lower than many other herbicides, due to its low application rate, it is applied to a large acreage and its usage continues to grow. In 1995, 27 percent of the corn acres was treated with dicamba, up from 16 percent in 1991. It was applied to 16 percent of the wheat in 1995, up from 10 percent in 1991.

Trifluralin is another important herbicide that has a long history of use. It is applied preplant and generally requires soil incorporation to control weeds as they germinate. Trifluralin is commonly used on cotton, soybean, potatoes and other vegetables. It is the mostly widely used herbicide on cotton with about 60 percent of the acres treated. Its use had been quite stable as no new cotton herbicides have become available and widely adopted in the past 15 to

20 years. Triflurlan use on soybeans has dropped since 1990 with the increase adoption of no-tillage systems where less capability exist for soil incorporation and with the availability of newer products. Imazethapyr, which first became available in the late 1980s, is now the most widely used soybean herbicide.

Alternative Measurement of Pesticide Use

Pesticide use in the United States, as traditionally reported, reached record levels in 1994. But, for assessing the human-health and environmental implications of pesticide use, weight of the materials may not be the most useful indicator of pesticide use. The amount of pesticides applied, measured in pounds or acres treated, fails to account for the wide variation both in toxicity per kilogram and persistence in the environment which characterizes the continuously changing array of more than 350 active ingredients that have been used in U.S. agricultural production in the last 40 years. Alternative measures of pesticide use, recently developed to measure pesticide use in terms of toxicity- and persistence-adjusted units, indicate a dramatically different trend in use over the 1964 to 1992 period.

Pesticide weight, as a measure of pesticide use, has two particularly notable drawbacks when used for evaluating the potential for harm to human health and the environment. First, pesticide active ingredients vary widely in terms of toxicity per unit of weight, irrespective of the scale used to measure toxicity.[8] Second, weight does not account for the persistence of the pesticide in the environment. The longer a pesticide ingredient remains active in the environment, the more potential there is for it to come in contact with unintended species. Persistence varies widely between active ingredients, but many modern pesticides have half-lives (the typical measure of persistence) in the range of 10 to 100 days.

A general perception is that, over time, there has been significant reduction in overall toxicity (especially in terms of measures relevant to long term

[8] Numerous measures of toxicity exist for individual pesticide active ingredients, including those designed to measure chronic and acute toxicity to humans, and toxicities to various avian, aquatic, and beneficial insect species. The relative toxicity of each pesticide ingredient varies depending upon which measure is used, and for a given measure, there is wide variation in toxicity among pesticide ingredients.

human health) of the substances applied in the environment as agricultural pesticides (Carlson and Wetzstein [1993] and Zalom and Fry [1992]). This perception is apparently based partly on the observation that many new pesticide compounds are applied at lower rates (pounds per acre) and are less persistent in the environment. In addition and as noted previously, an array of formerly widely used, but relatively highly toxic and persistent ingredients have been banned by the EPA. The information in this section provides a quantitative assessment of the accuracy of these perceptions. Adjustment factors are used to convert historical pesticide-use data (collected and reported in terms of pounds of active ingredients applied) into Persistence Units, Toxicity Units and Toxicity-Persistence Units. These terms are defined below and are more meaningful than pounds applied with respect to potential environmental and human health impacts. The weighing scheme adopted creates common denominators that reflect variation in toxicity and persistence among individual pesticide ingredients. Thus, the amounts of each pesticide active ingredient applied are aggregated via common units that are consistent across time, regions, pesticide types, toxicity, and persistence. This approach is consistent with other indexes designed to make assessments of aggregate changes in pesticide toxicity and persistence (e.g., Kovack et al. [1992] and Levitan et al. [1995]). The measures are converted to indexes to avoid the difficulties associated with measures that are not unit free. This is useful when the concern is with trends or the changes in toxicity and persistence relative to some base period. In these instances, the precise units associated with the measurement of persistence or toxicity are unimportant.

Persistence units (PERSIST) are based on soil half-life. This is length of time it takes for a pesticide to break down to half of its initial concentration. The soil half-life is highly dependent upon environmental conditions such as soil pH and climate. Organochlorines and some organophosphates tend to be very persistent in the environment. PERSIST units are created by calculating the half-life of one kilogram of each pesticide active ingredient. Multiplication of the index number for each pesticide active ingredient by pounds applied yields the total PERSIST units for each pesticide ingredient. The PERSIST units for each ingredient are then summed to obtain an aggregate measure of PERSIST. That is,

$$(1)\ PERSIST_t = (\sum_i (\delta_i\ \omega_{it})/\sum_i (\delta_i\ \omega_{i(base)}))$$

where PERSIST denotes the pesticide persistence index, δ_i denotes the half-life of one kilogram of pesticide i, ω_{it} denotes the number of pounds of active ingredient of pesticide i applied in period t, and $\omega_{i(base)}$ denotes the number of pounds of active ingredient of pesticide i applied in the base period. The summation is across all pesticides i.

Toxicity units (TUs) are based on indexes of the toxicity of each individual active ingredient. TUs are created by calculating the number of units (reference dose or LD50 - see below) contained in one kilogram of each pesticide active ingredient. Multiplication of the index number for each pesticide active ingredient by pounds applied yields the total TUs for each pesticide ingredient. The TUs for each ingredient are then summed to obtain an aggregate measure of TUs. That is,

(2) CHRONIC-TUs = $(\sum_i (\zeta_i\ \omega_{it})/\sum_i (\zeta_i\ \omega_{i(base)}))$

where CHRONIC-TUs denotes the pesticide chronic toxicity units index, ζ_i denotes the reference dose of one kilogram of pesticide i, ω_{it} denotes the number of pounds of active ingredient of pesticide i applied in period t, and $\omega_{i(base)}$ denotes the number of pounds of active ingredient of pesticide i applied in the base period. The summation is across all pesticides i. For acute toxicity,

(3) ACUTE-TUs = $(\sum_i (\lambda_i\ \omega_{it})/\sum_i (\lambda_i\ \omega_{i(base)}))$

where ACUTE-TUs denotes the pesticide acute toxicity units index, λ_i denotes the lethal dose of one kilogram of pesticide i, ω_{it} denotes the number of pounds of active ingredient of pesticide i applied in period t, and $\omega_{i(base)}$ denotes the number of pounds of active ingredient of pesticide i applied in the base period. The summation is across all pesticides i.

Toxicity-Persistence Units (TPUs) are based on further adjustment of the TUs to account for the combined toxicity and persistence of each individual active ingredient. The TPU index numbers are created by multiplying the toxicity index numbers (number of units of reference dose or LD50 contained in one kilogram of active ingredient) by the number of days (as measured by half-life) that an application of the active ingredient remains active in the environment. These Toxicity-Persistence index numbers are multiplied by the pounds of their respective active ingredient used and then summed to get aggregate TPUs. That is,

$$(4)\ \text{CHRONIC-TPUs} = (\textstyle\sum_i (\delta_i\, \zeta_i\, \omega_{it})/\sum_i (\delta_i\, \zeta_i\, \omega_{i(base)}))$$

where CHRONIC-TPUs denotes the pesticide chronic toxicity-persistence units index and the other terms are as previously defined. Acute toxicity-persistence units are defined as

$$(5)\ \text{ACUTE-TPUs} = (\textstyle\sum_i (\delta_i\, \lambda_i\, \omega_{it})/\sum_i (\delta_i\, \lambda_i\, \omega_{i(base)}))$$

where ACUTE-TPUs denotes the pesticide acute toxicity-persistence units index and the other terms are as previously defined.

Five separate measures of pesticide use are created for comparison to pounds of active ingredients. The first measures the persistence of pesticides in the environment. The next two measures (labeled Toxicity Units or TUs) adjust pounds applied for two alternative measures of pesticide toxicity. One of these measures adjusts pounds applied for chronic toxicity using an indicator of long term human health called the reference dose.[9] These units are labeled CHRONIC-TUs. The other measure adjusts pounds applied for acute pesticide toxicity using an indicator of acute mammalian toxicity called Oral LD50.[10] These units are labeled ACUTE-TUs. The acute measure is more relevant to one-time exposures while the chronic measure is relevant to daily exposure over an extended period.

The two toxicity measures can be further adjusted to account for persistence of individual pesticides in the environment. This adjustment, based on pesticide half-life, creates two additional measures called Toxicity-Persistence Units (TPUs). The terms CHRONIC-TPUs and ACUTE-TPUs are used to distinguish the TPU measures based on the reference dose from the one based on Oral LD50.

---

[9] The reference dose represents the maximum daily human exposure to that pesticide that results in no appreciable risk. The reference dose for each pesticide is determined from the no-observed-effect-level (NOEL) multiplied by a safety factor where NOEL is the maximum dose level (amount of pesticide/amount of body weight/day) at which no effects attributable to the pesticide under examination can be found.

[10] LD50 is used to measure the oral and dermal toxicity of a chemical and is expressed in terms of weight of the chemical per unit of body weight. It measures the amount of the toxicant necessary to kill 50 percent of the organisms being tested within a specified time period. The U.S. Environmental Protection Agency has identified the most toxic pesticides as being in Toxicity Class I and includes such pesticides as 2,4-D, methyl bromide, metam sodium, dicamba, terbufos, parathion, methomyl, and carbofuran.

Table 2.4. Index of relative use of pesticide active ingredients for selected years, comparing pounds and persistence, chronic toxicity, acute toxicity, and chronic and acute toxicity/persistence units.[1]

Pesticide type	Units	1964[2]	1966	1971	1992
Herbicides:					
	pounds	100	160	364	717
	PERSIST	100	198	493	1381
	CHRONIC-TUs	100	141	338	684
	ACUTE-TUs	100	136	246	321
	CHRONIC-TPUs	100	163	344	838
	ACUTE-TPUs	100	145	283	705
Insecticides:					
	pounds	100	95	107	57
	PERSIST	100	94	68	18
	CHRONIC-TUs	100	126	104	43
	ACUTE-TUs	100	99	168	149
	CHRONIC-TPUs	100	125	75	5
	ACUTE-TPUs	100	382	115	111
Fungicides:					
	pounds	100	98	139	159
	PERSIST	100	116	112	348
	CHRONIC-TUs	100	23	27	126
	ACUTE-TUs	100	361	377	509
	CHRONIC-TPUs	100	133	120	648
	ACUTE-TPUs	100	179	160	744
Other Pesticides:					
	pounds	100	85	112	404
	PERSIST	100	61	181	105
	CHRONIC-TUs	100	116	148	158
	ACUTE-TUs	100	95	142	327
	CHRONIC-TPUs	100	105	134	173
	ACUTE-TPUs	100	38	139	45
Total Pesticides:					
	pounds	100	109	171	264
	PERSIST	100	99	120	155
	CHRONIC-TUs	100	124	114	70
	ACUTE-TUs	100	100	168	155
	CHRONIC-TPUs	100	125	76	11
	ACUTE-TPUs	100	80	118	110

[1] Estimates are constructed for corn, soybeans, wheat, cotton, sorghum, rice peanuts, potatoes, other vegetables, and citrus.

[2] 1964=100.

Source: Economic Research Service [1997].

Aggregate pesticide use is compared in Table 2.4 for four points in time: 1964, 1966, 1971, and circa 1992. Comprehensive pesticide use data allowing for the computation of TUs and TPUs are available only for these years. Pesticide use is compared first on the basis of pounds of active ingredient applied, then on the basis of the two variations in the TUs and TPUs measures.

The TU and TPU indexes are simple measures related to the *potential* for exposure to chemicals with health effects. They do not reflect actual exposures. For example, these indicators do not consider how product formulation (e.g., liquid and granular forms and/or carrying agents) or application equipment have changed nor do they consider the proximity of humans who could be exposed. Such considerations would affect exposure and risk measures associated with chemical applications.

Table 2.4 shows that aggregate pesticide use in 1992 was 264 percent of the use in 1964, when measured by pounds of active ingredient. When measured by PERSIST, which adjusts for the half-life of the pesticides applied, the 1992 index is 55 percent greater than the 1964 value. When measured by CHRONIC-TUs, which adjusts for just toxicity based on the reference dose, the 1992 indicator is 30 percent below the 1964 level. Finally, when measured by CHRONIC-TPU, which adjusts for persistence and toxicity, the 1992 use is 89 percent below the 1964 use level. Clearly, the different series portray a dramatically different picture both of current use among pesticide types and of the change in pesticide use over time, with correspondingly different implications for human health and the environment. When pesticide use is measured in units that adjust for persistence and chronic human toxicity, it is clear that the character of current pesticide use is significantly different than it was in 1964, 1966 and 1971. Much of the reduction in PERSIST, CHRONIC-TUs and CHRONIC-TPUs since 1971 corresponds to the removal from the pesticide market of many organochlorine pesticides, such as aldrin, DDT, chlordane, and toxaphene.

Indicators of PERSIST, CHRONIC-TUs and CHRONIC-TPUs for herbicides and fungicides have increased substantially since 1964, while those for insecticides have declined in equally dramatic fashion. Insecticides account for such a large share of CHRONIC-TUs and CHRONIC-TPUs, however, that the increased volume of herbicides and fungicides did not offset the decline in insecticide PERSIST, CHRONIC-TUs and CHRONIC-TPUs. Insecticides account for more that 50 percent of the total CHRONIC-TUs and CHRONIC-TPUs and 75 percent of PERSIST currently. In 1964, insecticides accounted for 75 percent of total PERSIST, 80 percent of all CHRONIC-TUs and 97 percent of all CHRONIC-TPUs.

The picture is quite different when toxicity is defined in acute terms. Table 2.4 indicates that the ACUTE-TU indicator for pesticides is 55 percent greater than it was in 1964 and when measured by ACUTE-TPUs is 10 percent greater in 1992. The implication is that the amount of ACUTE-TUs and ACUTE-TPUs applied in the environment by use of the current array of pesticides is greater than in 1964. Herbicide and fungicide ACUTE-TPUs have increased dramatically (though neither accounts for a major percentage of total ACUTE-TPUs), while the insecticide measure has remained at roughly the same level. The aggregate insecticide ACUTE-TPUs is similar to 1964 levels, and has continued to account for slightly more than 91 percent of total ACUTE-TPUs. The insecticide ACUTE-TUs is currently above the 1964 level by more than 50 percent while it continues to account for approximately 94 percent of total ACUTE-TUs.

The fact that one kilogram of a pesticide is not necessarily equal to a kilogram of a different pesticide can be significant depending on how pesticide information is used. While pounds of pesticide material used is the most common method of measuring agricultural chemical use, the type of analysis undertaken will define what measure of chemical use is appropriate. Quantifying the risk from the exposure to pesticides, for example, typically requires weighing usage or residues by acute or chronic health and environmental toxicity coefficients and subsequently estimating human or environmental exposure to such hazards. The inferences one draws concerning pesticide use can vary substantially depending on the measure considered.

REFERENCES

Aspelin, A. *Pesticide Industry Sales and Usage: 1992 and 1993 Market Estimates.* Biological and Economic Analysis Division, Office of Pesticide Programs, US EPA, 733-K-94-001, June 1994.

Ball, E., and R. Nehring, "Productivity: Agriculture's Growth Engine," *Agricultural Outlook*, Economic Research Service, U.S. Department of Agriculture, May 1996.

Ball, E., J.C. Bureau, R. Nehring, and A. Somwaru, "Agricultural Productivity Revisited," *American Journal of Agricultural Economics*, forthcoming 1997.

Carson, G., "Long-Run Productivity of Pesticides," *American Journal of Agricultural Economics,* Vol. 59, No. 3 (1977).

Carlson, G., and E.N. Castle, "Economics of Pest Control," *Pest Control Strategies for the Future*, National Academy of Sciences, Washington, 1972.

Calson, G., and M. Wetzstein, "Pesticides and Pest Management," in *Agricultural and Environmental Resource Economics*, G. Carlson, D. Zilberman, and J. Miranowski, eds., Oxford University Press, New York, 1993.

Duffy, M., and M. Hanthorn. *Returns to Corn and Soybean Tillage Practices.* AER-508, USDA, ESCS, 1978.

Economic Research Service, *Agricultural Resources and Environmental Indicators*, U.S. Department of Agriculture, Economic Research Service, Washington, DC, July 1997.

Hawkins, D.E., W.F. Slife, and E. R. Swanson. "Economic Analysis of Herbicide Use in Various Crop Sequences," *Illinois Agricultural Economics*, Vol. 17, No. 1 (1977), pp. 8-13.

Headley, J.C. "Estimating the Productivity of Agricultural Pesticides," *Journal of Farm Economics,* Vol. 50 No. 1 (1968).

Huang, W.Y., L. Hansen, and N.D. Uri, "The Effects of the Timing of Nitrogen Fertilizer Application and Irrigation on Yield and Nitrogen Loss in Cotton Production," *Environment and Planning A*, Vol. 24 (1992), pp. 1449-1462.

Ibach, D.B., and M.S. Williams, "Economics of Fertilizer Use," *Fertilizer Technology and Use*, Edited by R.A. Olson, T.J. Army, J.J. Hanway and V.J. Kilmer, Soil Science Society of America, Madison, WI, 19971.

Kovach, J., C. Petzold, J. Degni, and J. Tette, "A Method to Measure the Environmental Impact of Pesticides," *New York's Food and Life Sciences Bulletin*, Vol. 139 (1992), pp. 33-47.

Levitan, L., I. Merwin, and J. Kovach, "Assessing the Relative Environmental Impacts of Agricultural Pesticides: The Quest for a Holistic Method," *Agriculture, Ecosystems, and Environment*, Vol. 55 (1995), pp. 153-168.

National Research Council, *Alternative Agriculture*, National Academy Press, Washington, 1989.

Pease, W. S., J. Liebman, D. Landy, and D. Albright, *Pesticide Use in California Strategies for Reducing Environmental Health Impacts*, Environmental Health Policy Program Report, University of California, Berkeley, CA, 1996.

U.S. Department of Agriculture, *Agricultural Statistics*, U.S. Government Printing Office, Washington, DC, 1998.

U.S. Environmental Protection Agency, Office of Prevention, Pesticides and Toxic Substances, Status of Chemicals in Special Review, EPA-738-A-95-001, Washington, D.C., May 1995.

Wallingford, G.W., "Cotton Fertilizer Efficiency - A Closer Look," *Farm Chemicals*, Vol. 155 (January 1992), pp. 58-60.

Wines, R.A., *Fertilizer in America: From Waste Recycling to Resource Exploitation*, Temple University Press, Philadelphia, Pa., 1985.

U.S. Food and Drug Administration, chemical use database, 1995.

CHAPTER 3

THE ECONOMIC AND ENVIRONMENTAL IMPLICATIONS OF ALTERNATIVE PRODUCTION PRACTICES

INTRODUCTION

While the known environmental and human health risks associated with the use of agricultural chemicals are, in general, relatively low (risks to farm workers being the chief exception), alternative production practices have been advocated as a means to reduce agricultural chemicals use and the risks associated with their use. The likely consequences of more extensive voluntary adoption of these alternative production practices is the subject of what follows.

Alternative production practices are typically thought of as being a systems approach to farming that are more responsive to natural cycles and biological interactions than conventional farming methods. For example, in alternative farming systems, farmers endeavor to integrate the beneficial aspects of biological interaction among crops, pests, and their predators into profitable agricultural systems. Some alternative farming practices are based on a number of accepted scientific principles and a substantial amount of empirical evidence. While it is not feasible to consider all such practices, some of the more widely touted are considered here. The specific mechanisms of many of these practices and their interactions, of course, need further study. In general, much is known about some of the components of alternative production practices, but not nearly enough is known about how the practices as a whole work.

Examples of practices or components of alternative production practices are listed below. Some of these practices are already part of conventional farming enterprises and are becoming more widely adopted. These include:

- Crop rotations that mitigate weed, disease, and insect problems; increase available soil nitrogen and reduce the need for synthetic fertilizers.
- Conservation tillage practices designed to slow soil erosion and conserve water through decreased soil disturbance.
- Genetic improvement of crops to resist pests and diseases and to use nutrients more effectively.

These alternative production practices are very general. To evaluate the impact that a more widespread adoption of such practices will have on input use, farm profitability, and the environment, it is necessary to be more specific as to the precise form the alternative production practices would take. To this end, the discussion will be limited to the first two production practices. Consideration of the last one, since it involves currently emerging technology, is properly relegated to a subsequent chapter.

CROP ROTATIONS

The Food Security Act of 1985 and the Food, Agriculture, Conservation, and Trade Act of 1990 strengthened the Federal Government's role in protecting soil and water quality. Besides increasing penalties for failing to comply with program provisions, the Act established programs offering incentives to remove land from production or to adopt alternative production practices to improve water quality and/or control erosion. Crop rotations and crop residue management systems are common practices recommended to reduce soil loss and to protect water resources from agricultural chemical contamination (National Research Council [1989]). Crop rotation is the successive planting of different crops in the same field. An example is corn followed by soybeans followed by oats followed by alfalfa. Rotations are the antithesis of continuous or monocropping which involves successively planting the same field with the same crop. Rotations typically range between 2 to 5 years in length but sometimes more and generally involve a farmer planting a portion of his cropland to each crop in the rotation. Rotations provide well documented economic and environmental benefits to agricultural producers (Heady and Jensen [1951], Heichel [1987], Kilkenny [1984], and Voss and Schrader [1984]).

Some of these benefits are endemic to any rotation while others depend on the crops planted, their sequencing, and length of the rotations, while yet others depend on the types of tillage, fertilization, and pest control practices employed in the rotation. Thus, crop rotations can increase yields, improve soil quality, reduce soil loss, conserve soil moisture, and reduce fertilizer and pesticide needs. Crop rotations, however, can potentially reduce net farm income when the acreage and frequency devoted to relatively highly profitable crops are replaced (rotated) with crops generating lower net farm income.

A substantial portion of the existing literature on crop rotations concerns the rotation effect (Heichel [1987] and Power [1987]). This rotation effect describes the fact that in most instances rotations will increase yields of a grain crop beyond yields achieved with continuous cropping under analogous conditions. The effect has been shown to exist whether rotations include nonleguminous or leguminous crops. In one study, corn following wheat, which is not a leguminous crop, produces greater yields than continuous corn when the same amount of fertilizer is applied (Power [1987]). In another study of 13 different cropping systems for corn, corn yields and profits were found to be greater when corn was rotated with other crops than when planted continuously (Helmers [1986]).

A number of factors are thought to contribute to the rotational effect, including increased soil moisture, pest control, and the availability of nutrients. It is generally agreed that the most important component of this effect is the insect and disease control benefits of rotations (Cook [1986]). The increase in soil organic matter may be the basis for the improved physical characteristics of soil observed in rotations which accounts for some yield increase. Certain deep-rooted leguminous and nonleguminous crops in rotations use soil nutrients from deep in the soil profile. In the process, these plants bring the nutrients to the surface, making them available to a subsequent shallow-rooted crop if crop residue is not removed.

The ability of legume crops to fix atmospheric nitrogen and supply soil nitrogen needs for subsequent crops is well documented. The plowdown of established alfalfa or other legumes can provide carryover nitrogen for a crop that requires high levels of nitrogen such as corn. Research has shown that soybeans can be managed to fix 90 percent of their nitrogen needs and provide a soil nitrogen credit of 20 pounds or more per acre for a subsequent crop (Heichel [1987]). Soybeans grown in rotation with corn, however, where soils are already rich in nitrogen have not been shown to fix significant amounts of nitrogen.

Crop rotations affect pest populations and can reduce the need for pesticides. Different crops often break pest cycles and prevent pest and disease organisms from building to damaging levels. Treatment for corn rootworm, the most common insecticide treatment on corn, normally only requires alternating another crop to reduce sufficiently rootworm survival rates to levels that do not require insecticide treatment. Hay and grass sod grown in rotation with corn, however, may increase the need for other corn insecticides to treat other pests.

Many crop rotations reduce soil loss and are an option for meeting conservation compliance on highly erodible land.[11] The growth of hay, small grain crops, or grass sod in rotation with conventionally tilled row crops reduces the soil's exposure to wind and water and decreases total soil loss and mitigates nitrate leaching and runoff. These rotations, however, are a desirable option for a farmer only when profitable markets exist or the conservation crops can be utilized by on-farm livestock enterprises.

Besides providing erosion control, small grains, hay, and grass sod are competitive with broad leaf weeds and may help control weed populations in subsequent crops thereby mitigating the need for herbicide use. These crops are usually harvested or can be cut before weeds reach maturity and produce seed for germination the following season. Weeds on prior idle acres or fallow land may be controlled by either cutting or tilling to reduce weed infestations the following year. Sometimes, herbicides are used to kill existing vegetation on idle land in lieu of mechanical methods.

As noted, crop rotation is frequently key to a sustainable agricultural production system and can reduce the need for fertilizer and pesticides. Fewer pesticides may be needed when rotations break pest cycles or reduce infestation levels. The use of herbicides in 1995 varied little between different corn rotations, but continuous corn more frequently used insecticides, especially to control corn rootworm by 61.1 percent for continuous corn versus 17.2 per-

---

[11] The Food Security Act of 1985 targeted highly erodible land under the conservation compliance provision. This provisions stipulated that farmers of such land who did not implement approved conservation plans would lose eligibility for U.S. Department of Agriculture program benefits. Under conservation compliance, farmers with highly erodible cropland had to have a plan in place by January 1, 1990 and that plan implemented by January 1, 1995. Many different conservation tillage practices were acceptable.

cent for a three-year rotation (Economic Research Service [1997]). Rootworm larvae populations most frequently reach damaging infestations when corn follows corn. Alternating another crop with corn usually eliminates the need for insecticide treatment for corn rootworm.

Both herbicide and insecticide use were higher in 1995 on continuous cotton than cotton in rotation. While there was little difference in the proportion of acres treated, the annual per-acre treatments were greater on continuous cotton. The yearly quantity of pesticides applied per acre to continuous cotton was approximately 50 percent higher that to cotton in rotation with other row crops. On average, about one additional herbicide and one additional insecticide ingredient were used with continuous cotton. Continuous cotton received an average of 7.1 pesticide treatments and 6.8 different pesticide ingredients compared with 5.1 treatments and 5 ingredients for other rotations (Economic Research Service [1997]).

Relative to cotton and corn, soybeans and wheat used fewer pesticides in 1995. Nearly all soybeans were treated with a herbicide and there was little difference in herbicide use between rotations. Soybeans were rarely treated with an insecticide. Nearly 40 percent of wheat received no pesticides and on the treated acres, the average number of products and application rates were much lower than for either cotton or corn. For example, the average application rate for herbicides and insecticides, respectively, for wheat was 0.31 pounds per acre and 0.35 pounds per acre while it was 2.64 pounds per acre and 0.89 pounds per acre for corn (Economic Research Service [1997]).

Unlike continuous cropping of corn and cotton, continuous wheat received pesticides less frequently than wheat in rotation. Only about half of the acres in continuous wheat were treated with a herbicide in 1995. About two-thirds of all wheat fallowed or in rotation with other crops was treated with herbicides and the average application rate was about one-third higher than for continuous wheat (Economic Research Service [1997]).

For fertilizer, when the previous crop was hay or pasture, lower nitrogen applications were reported for corn - 85 pounds per acre in rotation with hay or pasture versus 129 pounds when the land was in continuous row crops (Economic Research Service [1997]).

A few studies exist that evaluate the economic and environmental effects of adopting crop rotations in place of continuously planting a crop. Huang and Uri [1992] look at reducing the nitrogen available for potential leaching into the groundwater under alternative production practices. Using a nitrogen mass balance approach which accounts for all nitrogen applied and available for plant uptake and all nitrogen removed after the crop is harvested in a nonlinear mathematical programming framework, models of continuous cropping and crop rotations where Iowa corn is rotated with soybeans and meadow are developed. The excess nitrogen applied that is available for potential leaching varies among different crop rotations. Continuous planting of corn over a three-year rotation has a large excess nitrogen fertilizer application rate relative to a rotation involving corn and soybeans and a rotation involving corn and alfalfa (meadow). In fact, excess nitrogen is nearly twice as great under continuous planting of corn than it is under any type of rotation. An analysis of the loss of net farm income is performed in the context of a number of policy options including taxing the use of nitrogen fertilizer and taxing the output of the commodity on which nitrogen fertilizer is used. The authors find that moving from monocropping to some sort of rotation results in a reduction of net farm income. Soybeans rotated with corn, however, yield only minimal adverse effects.

In another study, Huang and Uri [1991] use a mixed integer, nonlinear mathematical programming model of agricultural production to assess the impact on the nitrogen fertilizer application rate of moving from the continuous planting of corn to a crop rotation involving corn and soybeans. The data are taken from the field experiment studies conducted at the Iowa State University Experiment Station at Kanawha, Iowa. The results are acknowledged to be static and partial equilibrium in nature. The results which optimize net farm income indicate that for a six year rotation, the nitrogen fertilizer application rate will average about 5 to 10 percent higher annually for the continuous planting of corn and result in an average corn yield of about 7 to 8 percent less than that for a corn-soybean rotation.

In yet another study, Huang and Uri [1994] assess the impact of crop rotations on reducing the leaching potential of nitrogen fertilizer applied. A dynamic optimization model is used. A farmer's decision with regard to the amount of nitrogen fertilizer applied to any cropping pattern is formally modeled. The assumed objective is the maximization of net per acre return. The continuous planting of corn is found consistently to yield the highest return when compared to corn-soybean and corn-meadow (alfalfa) rotations. When a farmer is required to limit the amount of nitrogen fertilizer available for potential leaching, however, a soybean-corn rotation proves optimal.

Wu et al. [1996] look at the effect of crop rotations on the potential nitrogen available for runoff and leaching. An empirical model is developed for all National Resources Inventory sites in the Midwest and Northern Plains of the United States. The factor precipitating a change from monocropping to crop rotations is assumed to be the passage of the Federal Agriculture Improvement and Reform Act of 1996. Based on studies by Hennessy et al. [1995] and Williams et al. [1990], it is presumed that in the Corn Belt and Lake States, continuous planting of corn will be switched to corn-soybeans rotations, and continuous wheat and continuous sorghum will be switched to a wheat-sorghum rotation without farm program constraints. In the Northern Plains, continuous corn will be switched to a corn-soybeans rotation, and continuous wheat and continuous sorghum will be switched to the wheat-sorghum-fallow rotation. There is some change from monocropping to crop rotations. Approximately 20 percent of available cropland is switched from single cropping to rotations. Nitrogen use falls as a consequence with the result that nitrate leaching declines to in excess of twenty percent in some regions. In Nebraska, for example, nitrate leaching declines by 23.1 percent when continuous cropping is replaced by crop rotations. Finally, net farm income increases when, in the absence of a farm program, continuous cropping is replaced by crop rotations.

Conservation Tillage

Farmers have adopted a range of conservation tillage practices over the past four decades designed to maintain or enhance soil characteristics (Bruce et al. [1995], Thorne and Thorne [1979] and Wood et al. [1991]). Most of these practices have been designed to affect soil characteristics that slow soil erosion and influence the movement of water in and through the soil (e.g., structure, organic matter content, and soil microbial populations) (Office of Technology Assessment [1990]). For example, conservation tillage practices[12] that leave substantial amounts of crop residue evenly distributed over the soil surface defend against the potential of rainfall's kinetic energy to generate sediment and increase water runoff. Several field studies (e.g., Baker and Johnson [1979], Glenn and Angle [1987], Hall et al. [1984], and Sander et al. [1989]) conducted under natural rainfall on highly erodible land (14 percent

[12] Conservation tillage is defined to be any tillage or planting system that maintains at least 30 percent of the soil covered by residue after planting to reduce soil erosion by water; or where soil erosion by wind is the primary concern, maintains at least 1,000 pounds (per acre) of flat, small grain residue equivalent on the surface during the critical wind erosion period (Bull [1993]).

slope) have compared erosion rates between tillage systems. Compared with moldboard plowing, no-tillage[13] generally reduced soil erosion by more than 90 percent while mulch tillage and ridge tillage[14] reduced soil erosion by 50 percent or more. Specifically, Hall et al. [1984] report that when compared with conventional tillage, no-tillage reduced runoff by 86.3 to 98.7 percent, soil losses by 96.7 to 100 percent, and the herbicide cyanazine losses by 84.9 to 99.4 percent. Glenn and Angle [1987] report a 27 percent less total runoff of water and the herbicides atrazine and simazine with no-tillage versus conventional tillage systems.

Increased surface residues also filter out and trap sediment and sediment absorbed agricultural chemicals (pesticides and fertilizer) and result in a cleaner runoff (Onstad and Voohees [1987]). The increase in organic matter associated with crop residue management intercepts the chemicals and holds them in place until they are used by the crop or degrade into inert components (Dick and Daniel [1987] and Wagnet [1987]). The presence of increased crop residue typically reduces the volume of contaminants entering surface water by constraining runoff (including dissolved chemicals and sediment) and enhancing filtration (Baker [1987], Fawcett [1987], and Wauchope [1987]).

Enhanced water infiltration associated with greater surface residue provides additional soil moisture to benefit crops during low rainfall periods but raises concerns about potential leaching of nitrates and pesticides to shallow groundwater (Baker [1987] and Wauchope [1987]). Evidence suggests, however, that under normal climatic and hydrologic conditions, conservation tillage practices such as no-tillage, mulch tillage, and ridge tillage are no more likely to degrade groundwater or surface water quality than conventional tillage practices (Baker [1980], Edwards et al., [1993], and Fawcett et al. [1994]).

While conservation tillage often requires different herbicide treatments, all tillage practices use a broad spectrum of herbicides for weed control. Many

[13] Under a no till cropping practice, the soil is left undisturbed before planting. Planting is completed in a narrow seedbed or slot created by a planter or drill. Weeds are controlled primarily with herbicides and/or after planting cultivation (Bull [1993]).

[14] With mulch tillage, the total surface is disturbed by tillage before planting. Tillage tools such as chisels or field cultivators (disks, sweeps, or blades) are used. Weeds are controlled with herbicides and/or cultivation. Under the ridge tillage production practice, the soil is left undisturbed before planting. Planting is performed on ridges in a seedbed prepared with sweeps, disks, or other row cleaners. Residue is left on the surface between ridges. Weeds are controlled with herbicides and/or cultivation. Ridges are rebuilt during the growing season with cultivation for weed control (Bull [1993]).

factors, including the type of chemical applied, application method and timing, soil properties, climatic conditions, and the crop residue effects on compaction and macropores can affect the fate of applied nutrients and pesticides (Bull et al. [1993]).

Conservation tillage practices clearly do have a beneficial environmental impact. Yet, there is a lingering question as to their overall efficacy in reducing the impact of agricultural production on the environment. This is because the nature of the changes in agricultural chemical use required by conservation tillage have not yet been comprehensively investigated (Kellogg et al. [1994], National Research Council [1989, 1993]). What is known is that conventional moldboard plowing, in contrast to conservation tillage, contributes to pest control by destroying some perennial weeds, disrupting the life cycle of some insect pests, and burying disease inoculum (Holland and Coleman [1987]). Additionally, conventional tillage creates more bacterial activity and has a "boom-and-bust" effect on nutrient cycling processes while no till or other conservation tillage provides a slower but more even rate of nutrient release (Heichel [1987]).

Changing the tillage system may have economic effects that are a direct result of a number of factors including site-specific soil characteristics, local climatic conditions, cropping patterns, and other attributes of the overall farming operation. Major costs directly affected by the choice of tillage practice are machinery and herbicides. In general, decreasing the intensity of tillage operations or reducing the number of operations results in lower machinery, fuel, and labor costs. These cost savings may be offset by potential increase in agricultural chemical costs depending on the herbicides selected for weed control and the fertilizer required to obtain optimal yields (Siemens and Doster [1992]).

There are indirect effects from a change in tillage systems. These include the impact on the rate of soil erosion and qualitative characteristics of the soil and their long run influence on soil productivity.

It has been argued that a farmer adopts a conservation tillage system not only to conserve soil and mitigate any adverse impacts that agricultural production has on the environment but also to save time and money (Gadsby et al. [1987]). Data show that for corn and soybeans, the number of labor hours devoted to tillage operations range from 0.4 to 0.6 per acre for conventional tillage and from 0.1 to 0.3 for conservation tillage systems (Monson and Wollenhaupt [1995]).

A major advantage from the adoption of conservation tillage involves fewer trips over the field that reduce fuel and maintenance costs. Fuel costs, like labor costs, are estimated to drop by up to 60 percent per acre. Tillage equipment lasts longer when used less intensively and extending machinery life reduces ownership costs (Monson and Wollenhaupt [1995]).

Conservation tillage systems require improved fertilizer management. In some instances, increased application of specific nutrients may be necessary and specialized equipment required for proper fertilizer placement, thereby contributing to higher costs. Moreover, improved nutrient management techniques involving the proper timing and placement of the intended crop's nutrient requirements are necessary with all tillage systems to reduce the potential for nutrient losses to the environment. These improvements usually are associated with an enhancement in environmental quality (National Research Council [1993]).

The effect of conservation tillage on pesticide use is uneven. Users of no-tillage applied more herbicides and less insecticides per acre than other tillage systems. Mulch-till systems, however, often applied less herbicide than when conventional tillage is used (Economic Research Service [1997]).

The impact of conservation tillage practices on yield varies with location, soil type, climatic conditions, level of management, and the crop produced (Foltz et al. [1995] and Fox et al. [1991]). For example, there are no apparent yield differences for corn, cotton, and soybeans in the Southeast using no-tillage versus conventional tillage on well drained to moderately well drained soils. Long run studies show slightly higher no-tillage yields and experienced no-tillage farmers claim greater yields from increased soil organic matter (Hudson and Bradley [1995]). In some areas of the Northern Great Plains, the benefit of improved moisture retention associated with conservation tillage results in higher crop yields and permits a change in the cropping pattern to reduce the frequency of moisture conserving fallow periods (Clark et al [1994]).

One study looked at the economic effects of six different tillage systems on three soil types for continuous corn and a corn/soybeans rotation (Doster et al. [1994]). It concludes that net returns are greater for a corn/soybeans rotation relative to continuous corn for all tillage systems for the three soil types. On highly erodible soil, however, no-tillage generates the highest net-returns while ridge-till yields the greatest returns for the other soil types.

Another study has examined the environmental consequences of a more intensive adoption of conservation tillage. Lakshminarayan and Babcock [1996] look at the effects of voluntary replacement of 50 percent of conventional moldboard tillage with conservation tillage. An empirical model is developed for all National Resources Inventory primary sampling unit (PSU) sites in the Midwest and Northern Plains of the United States. It reflects the highly uncertain nature of the production costs associated with the adoption of conservation tillage. Farmers who successfully switch from conventional tillage to conservation tillage in general find their variable costs of production lower even though they incur some costs in making the requisite adjustments including purchasing conservation tillage equipment and acquiring technical assistance. The cost to a farmer in switching from conventional tillage to conservation tillage is estimated to be $1.76 per acre. No estimate is provided for the average savings on variable costs of production after conservation tillage is adopted. The environmental benefits come primarily in the form of reduced soil erosion. No assessment of changes in fertilizer or pesticides use is offered.

Lin et al. [1995] assess the factors that influence herbicide use in corn production in the North Central region of the United States. The study uses data from the Cropping Practices Survey for corn production between 1990 and 1992 and specifies a single equation factor demand function for herbicides. The results suggest that farmers who switch from conventional tillage with the moldboard plow to conservation tillage do increase herbicide use. Conventional tillage without the moldboard plow uses herbicides at a rate similar to conservation tillage. Among the various types of conservation tillage, no-tillage uses the most herbicides. Conventional tillage with moldboard plow among all types of tillage practices uses the least amount of herbicides.

THE ROLE OF VOLUNTARY PROGRAMS IN REDUCING AGRICULTURAL CHEMICAL USE

The adoption of either crop rotations or conservation tillage by a farmer is voluntary. While the environmental benefits are potentially sizeable, the economic benefits are less transparent. At issue then is how to induce a farmer to voluntarily adopt an alternative production practice that might not be in his best interest. Programs aimed at reducing agricultural chemical use and the attendant nonpoint source pollution traditionally have relied on voluntary programs. These programs frequently couple technical assistance with a subsidy covering a portion of the recommended production practice's cost. (The following chapter discusses this issue in greater detail.) A farmer, however, is unlikely to adopt voluntary pollution-reducing practices in the absence of a subsidy covering the costs of the practice plus any revenue reduction (Norton et al. [1994]).

Voluntary programs coupled with subsidies are not free of shortcomings, but in certain cases voluntary programs may be very effective that might be expected because farming possesses a unique structure. In farming, the owner/operator and the owner/operator's family frequently reside on the farmsite. In fact, in 1993 99.1 percent of all farms were sole proprietorships, partnerships, or family held corporations (U.S. Department of Agriculture [1994]). Consequently, the owner and decision-maker on a farm may bear a portion of the pollution costs of their production decisions and thus have an incentive to reduce that pollution. For example, fertilizer and pesticides can leach into the farmer's own well water and endanger family health in addition to polluting water off the farm. The farmer may be willing to adopt pollution-reducing technologies without full compensation for a loss in net farm income if the technology also increases on-farm environmental quality. Norton et al. [1994] develop the conditions under which a farmer would be willing to voluntarily adopt a pollution reducing alternative production practice. The essential condition is that government subsidy payments must exceed a farmer's willingness to accept (a subjective measure based on the farmer's utility) a decrease in net farm income.

There are alternatives to voluntary programs that are not coupled with subsidies including regulation and taxes on the purchase of pesticides and fertilizer. This is the subject of the following chapter.

REFERENCES

Baker, J.L., "Hydrologic Effects of Conservation Tillage and Their Importance to Water Quality," in T. Logan, J. Davidson, J. Baker, and M. Overcash (eds.), *Effects of Conservation Tillage on Groundwater Quality: Nitrates and Pesticides*, Lewis Publishers, Chelsea, MI, 1987, pp. 113-124.

Baker, J., and H. Johnson, "The Effect of Tillage Systems on Pesticides in Runoff from Small Watersheds," *Transactions of the American Society of Agricultural Engineers*, Vol. 22 (1979), pp. 554-559.

Bosch, D., Z. Cook, and K. Fuglie, "Voluntary Versus Mandatory Agricultural Policies to Protect Water Quality: Adoption of Nitrogen Testing in Nebraska," *Review of Agricultural Economics*, Vol. 17 (1995), pp. 13-24.

Bruce, R., G. Langdale, L. West, and W. Miller, "Surface Soil Degradation and Soil Productivity Restoration and Maintenance," *Soil Science Society of America Journal*, Vol. 59 (1995), pp. 654-660.

Bull, L., *Residue and Tillage Systems for Field Crops*, SR AEGS 9310, Resources and Technology Division, Economic Research Service, U.S. Department of Agriculture, Washington, July 1993.

Bull, L., H. Delvo, C. Sandretto, and W. Lindamood, "Analysis of Pesticide Use by Tillage System," *Agricultural Resources: Inputs Situation and Outlook Report*, U.S. Department of Agriculture, Economic Research Service, Washington, DC, 1993.

Clark, R., J. Johnson, and J. Brundson, "Economics of Residue Management, Northern Great Plains Region," in *Crop Residue Management to Reduce Soil Erosion and Improve Soil Quality: North Central Region*, Agricultural Research Service, U.S. Department of Agriculture, Washington, DC, 1995.

Cook, R.J., "Wheat Management Systems in the Pacific Northwest," *Plant Disease*, Vol. 70 (1986), pp. 894-898.

Dick, W.A., and T.C. Daniel, "Soil Chemical and Biological Properties as Affected by Conservation Tillage: Environmental Implications," in T. Logan, J. Davidson, J. Baker, and M. Overcash (eds.), *Effects of Conservation Tillage on Groundwater Quality: Nitrates and Pesticides*, Lewis Publishers, Chelsea, MI, 1987, pp. 315-339.

Economic Research Service, *Agricultural Resources and Environmental Indicators*, U.S. Department of Agriculture, Washington, DC, July 1997.

Edwards, W., M. Shipitalo, L. Owens, and W. Dick, "Factors Affecting Preferential Flow of Water and Atrazine Through Earthworm Burrows under Continuous No-Till Corn," *Journal of Environmental Quality*, Vol. 22 (1993), pp. 225-241.

Fawcett, R.S., "Overview of Pest Management for Conservation Tillage Systems," in T. Logan, J. Davidson, J. Baker, and M. Overcash (eds.), *Effects of Conservation Tillage on Groundwater Quality: Nitrates and Pesticides*, Lewis Publishers, Chelsea, MI, 1987, pp. 6-19.

Foltz, J., J. Lee, M. Martin, and P. Preckel, "Multiattribute Assessment of Alternative Cropping Systems," *American Journal of Agricultural Economics*, Vol. 77 (1995), pp. 408-420.

Fox, G., A. Weersink, G. Sarwar, S. Duff, and B. Deen, "Comparative Economics of Alternative Agricultural Production Systems: A Review," *Northeastern Journal of Agricultural and Resource Economics*, Vol. 30 (1991), pp. 43-52.

Fuglie, K., and D. Bosch, "Economic and Environmental Implications of Soil Nitrogen Testing: A Switching Regression Analysis," *American Journal of Agricultural Economics*, Vol. 77 (1995), pp. 891-900.

Gadsby, D., R. Magleby, and C. Sandretto, "Why Practice Conservation," *Soil and Water Conservation News*, Vol. 8 (1987), pp. 3-6.

Glenn, S., and J.S. Angle, "Atrazine and Simazine in Runoff from Conventional and No-Till Corn Watersheds," *Agriculture, Ecosystems, and Environment*, Vol. 18 (1987), pp. 273-280.

Goldstein, W., and D. Young, "An Agronomic and Economic Comparison of a Conventional and Low Input Cropping System in the Palouse," *American Journal of Alternative Agriculture*, Vol. 11 (Spring 1987), pp. 51-56.

Hall, J.K., L. Hartwig, and L. Hoffman, "Cyanazine Losses in Runoff from No-Tillage Corn in "Living" and Dead Mulches vs. Unmulched, Conventional Tillage," *Journal of Environmental Quality*, Vol. 13 (1984), pp. 105-110.

Heady, E., and H. Jensen, *The Economics of Crop Rotations and Land Use: A Fundamental Study in Efficiency with Emphasis on Economic Balance of Forage and Grain Crops*, Agricultural Experiment Station, Iowa State University, Ames, IA, 1951.

Heichel, G., "Legumes as a Source of Nitrogen in Conservation Tillage Systems," in *The Role of Legumes in Conservation Tillage Systems*, J.F. Power (ed.), Soil Conservation Society of America, Ankeny, IA, 1987.

Helmers, G., "An Economic Analysis of Alternative Cropping Systems for East Central Nebraska," *American Journal of Alternative Agriculture*, Vol. 1 (1986), pp. 15-22.

Hennessy, D., B.A. Babcock, and D. Hayes, *The Budgetary and Resource Allocation Effects of Revenue Assurance*, Working Paper 95-WP130, Center for Agriculture and Rural Development, Iowa State University, Ames, IA, February 1995.

Holland, E.A., and D.C. Coleman, "Litter Placement Effects on Microbial and Organic Matter Dynamics in an Agroecosystem," *Ecology*, Vol. 68 (1987), pp. 425-433.

Huang, W., L. Hansen, and N. Uri, "The Effects of the Timing of Nitrogen Fertilizer Application and Irrigation on Yield and Nitrogen Loss," *Environment and Planning*, Vol. 24 (1992), pp. 1449-1462.

Huang, W., and N.D. Uri, "An Analysis of the Marginal Value of Cropland and Nitrogen Fertilizer Use Under Alternative Farm Programs," *Agricultural Systems*, Vol. 35 (1991), pp. 433-453.

Huang, W., and N.D. Uri, "An Assessment of Alternative Agricultural Policies to Reduce Nitrogen Fertilizer Use," *Ecological Economics*, Vol. 5 (1992), pp. 213-234.

Huang, W., and N.D. Uri, "The Effect of Farming Practices on Reducing Excess Nitrogen Fertilizer Use," *Water, Air, and Soil Pollution*, Vol. 77 (1994), pp. 79-95.

Huang, W., L. Hansen, N.D. Uri, "The Application Timing of Nitrogen Fertilizer," *Water, Air, and Soil Pollution*, Vol. 73 (1994), pp. 189-211.

Hudson, E., and J. Bradley, "Economics of Surface Residue Management, Southeaster Region," in *Crop Residue Management to Reduce Soil Erosion and Improve Soil Quality: North Central Region*, Agricultural Research Service, U.S. Department of Agriculture, Washington, DC, 1995.

Kellogg, R., M. Maizel, and D. Goss, *Agricultural Chemical Use and Groundwater Quality: Where Are the Potential Problems*, National Center for Resource Innovations, Washington, DC, 1994.

Kilkenny, *An Economic Assessment of Biological Nitrogen Fixation in a Farming System of Southeast Minnesota*, M.S. thesis, University of Minnesota, St. Paul, MN, 1984.

Killorn, R., and D. Zourarakis, "Nitrogen Fertilizer Management Effect on Corn Grain Yield and Nitrogen Uptake," *Journal of Production Agriculture*, Vol. 5 (1992), pp. 142-148.

Lakshminarayan, P.G., and B. Babcock, *Temporal and Spatial Evaluation of Soil Conservation Policies*, Center for Agriculture and Rural Development, Iowa State University, Ames, IA, 1996.

Legg, T., J. Fletcher, and K. Easter, "Nitrogen Budgets and Economic Efficiency: A Case Study of Southeastern Minnesota," *Journal of Production Agriculture*, Vol. 2 (1989), pp. 110-116.

Lin, B., H. Taylor, H. Delvo, and L. Bull, "Factors Influencing Herbicide Use in Corn Production in the North Central Region," *Review of Agricultural Economics*, Vol. 17 (1995), pp. 159-169.

Monson, M., and N. Wollenhaupt, "Residue Management: Does It Pay? North Central Region," in *Crop Residue Management to Reduce Soil Erosion and Improve Soil Quality: North Central Region*, Agricultural Research Service, U.S. Department of Agriculture, Washington, DC, 1995.

National Research Council, *Alternative Agriculture*, National Academy of Sciences, Washington, 1989.

National Research Council, *Soil and Water Quality*, National Academy Press, Washington, DC, 1993.

Norton, N., T. Phipps, and J. Fletcher, "Role of Voluntary Programs in Agricultural Nonpoint Pollution Policy," *Contemporary Economic Policy*, Vol. 12 (1994), pp. 113-121.

Office of Technology Assessment, *Beneath the Bottom Line*, Congress of the United States, Washington, DC, November 1990.

Olson, R., "Nitrogen Problems," in *Plant Nutrient Use and the Environment*, Fertilizer Institute, Washington, DC, 1985, pp. 115-138.

Onstad, C., and W. Voorhees, "Hydrologic Soil Parameters Affected by Tillage," in T. Logan, J. Davidson, J. Baker, and M. Overcash (eds.), *Effects of Conservation Tillage on Groundwater Quality: Nitrates and Pesticides*, Lewis Publishers, Chelsea, MI, 1987, pp. 274-291.

Peterson, G., and W. Frye, "Fertilizer Nitrogen Management," in *Nitrogen Management and Groundwater Protection*, R. Follett (ed.), Elsevier Scientific Publishers, Amsterdam, 1989, pp. 183-220.

Peterson, G., and M. Russelle, "Alfalfa and the Nitrogen Cycle in the Corn Belt," *Journal of Soil and Water Conservation*, Vol. 46 (1991), pp. 229-235.

Porter, K., and M. Stimman, *Protecting Groundwater: A Guide for the Pesticide User*, Cornell University Cooperative Extension, Ithaca, 1988.

Power, J.F., "Legumes: Their Potential Role in Agricultural Production," *American Journal of Alternative Agriculture*, Vol. 2 (1987), pp. 69-73.

Sander, K., W. Witt, and M. Barrett, "Movement of Triazine Herbicides in Conventional and Conservation Tillage Systems," in D.L. Weigmann (ed.), *Pesticides in Terrestrial and Aquatic Environments*, Virginia Water Resources Center, Virginia Polytechnic Institute and State University, Blacksburg, pp. 378-382.

Siemans, J., and D.H. Doster, "Costs and Returns," in *Conservation Tillage Systems and Management, Crop Residue Management with No-Till, Ridge-Till, and Mulch Till*, Agricultural and Biosystems Engineering Department, Iowa States University, Ames, IA, 1992.

Thorne, D.W., and M.D. Thorne, *Soil, Water and Crop Production*, AVI Publishing Company, Inc., Westport, Connecticut, 1979.

U.S. Department of Agriculture, *Economic Indicators of the Farm Sector: National Financial Summary, 1993*, Economic Research Service, U.S. Department of Agriculture, Washington, DC, December 1994.

Voss, R.D., and W.D. Schrader, "Rotational Effects and Legume Sources of Nitrogen for Corn," in *Organic Farming: Current Technology and Its Role in a Sustainable Agriculture*, D. Bezdicek and J.F. Power (eds.), American Society of Agronomy, Madison, WI, 1984.

Wagnet, R.J., "Processes Influencing Pesticide Loss with Water under Conservation Tillage," in T. Logan, J. Davidson, J. Baker, and M. Overcash (eds.), *Effects of Conservation Tillage on Groundwater Quality: Nitrates and Pesticides*, Lewis Publishers, Chelsea, MI, 1987, pp. 189-200.

Williams, J.R., R.V. Llewelyn, and G.A. Barnaby, "Risk Analysis of Tillage Alternatives with Government Programs," *American Journal of Agricultural Economics*, Vol. 72 (1990), pp. 172-181.

Wood, C., J. Edwards, and C. Cummins, "Tillage and Crop Rotation Effects on Soil Organic Matter in a Typic Hadudult of Northern Alabama," *Journal of Sustainable Agriculture*, Vol. 2 (1991), pp. 31-41.

Wu., J., P.G. Lakshminarayan, and B. Babcock, *Impacts of Agricultural Practices and Policies on Potential Nitrate Water Pollution in the Midwest and Northern Plains of the United States*, Center for Agriculture and Rural Development, Iowa State University, Ames, IA, 1996.

CHAPTER 4

THE INFLUENCE OF GOVERNMENT POLICY ON THE CHOICE OF PRODUCTION PRACTICES AND CHEMICAL USE

INTRODUCTION

Government influence on the choice of agricultural production practices and the attendant use of chemicals has a variety of forms. Before exploring these alternatives, it is important to understand the rationale for government intervention: externalities arising from the interaction between the agricultural sector and the rest of society.

Externalities exist in situations where the activities of an economic agent affect the preferences, consumption, or technology of another economic agent (Buchanan and Stubblebine [1962]). Consider the application by a farmer of pesticides that run off into surface drinking water supplies and are ingested by individuals. Drinking water with high concentrations of pesticides has suspected risks and associated costs to human health. This is an example of a negative externality because the action of the farmer adversely affects the welfare of consumers.

The absence of externalities is one of the conditions required for competitive markets to achieve an efficient allocation of resources. This is not meant to imply, however, that the presence of an externality requires government intervention. In many situations the involved parties may negotiate a solution that will address the externality problem and result in an efficient resource allocation. For example, restricting pesticide spraying during certain times to minimize community exposure to drifting pesticides can be the result of a voluntary agreement between a farmer and the residents surrounding the farmer's cropland.

There are, however, externalities where the interaction between private parties does not lead to an efficient allocation of resources. Government intervention may be considered in these instances even though there is no guarantee that intervention will lead to enhanced efficiency. Such situations are referred to as externality problems or market failure. Government intervention can take a variety of forms including regulation, taxes, subsidies, and educational and technical assistance.

There are other situations where intervention is justified on the basis of distributional equity considerations. Even if an efficient resource allocation could be obtained through private or public approaches, the solution could be suboptimal from society's perspective if it results in inequities in terms of income distribution or the burden of regulation. Because distributional inequity is so highly subjective, however, little discussion will be devoted to it in what follows.

As previously noted, externalities play a central role in the economics of the interaction between the agricultural sector and the stock of natural resources. To mitigate the impact of externalities, a number of policy options are available to the government. These policy options in general have the potential to impact the production practices adopted by farmers and the use of agricultural chemicals. In what follows, these policy options will be explored along with their implications for the economy and the environment.

REGULATION

Direct government involvement in the market via regulation is one way that has been advocated to address the externalities arising from the use of agricultural chemicals. Regulation can take a variety of forms ranging from restrictions on the amount of an input that can be used to outright banning the use of an input because the externality resulting from its use is so egregious. It is generally agreed in the latter case that direct control, or a ban, is the most desirable form of government intervention in the market. That is, when the desired emission level is zero, banning the use of the input is the optimal policy option. A second situation when direct control is best occurs when rapid or temporary variation in pollution levels results as a consequence of, for example, changing weather patterns. Thus, for example, limiting the application of fertilizer during the spring rainy season could be used to control runoff into surface water and leaching into groundwater (Fisher [1981]). As an alternative to an outright ban, other regulatory restraints have been placed on input use.

Because of the potential human health and environmental risks associated with the use of chemical pesticides and fertilizer in agriculture, Congress has enacted several statutes designed to minimize these hazards while permitting the beneficial uses of these chemicals. Pesticide regulation in its modern form began with the enactment of the Federal Insecticide, Fungicide, and Rodenticide Act (FIFRA) in 1948. Under this mandate, Congress required that all chemicals for sale in interstate commerce be registered against the manufacturers' claims of effectiveness. The law also required manufacturers to indicate pesticide toxicity on the label. Congress amended FIFRA in 1954, 1959, and 1964, but, in practice, pesticide regulation by 1970 meant efficacy testing and labeling for acute (short-term) toxicity. Pesticide regulation passed into a new phase with the 1972 amendment to FIFRA and the transfer of regulatory jurisdiction to the EPA. Under this new regulatory regime, Congress gave the EPA the responsibility of re-registering existing pesticides, examining the effects of pesticides on fish and wildlife, and evaluating acute and chronic toxicity. In a 1988 amendment to FIFRA, pesticide producers were required to demonstrate, within 9 years, that all pesticides registered before November 1984 meet current standards (Ollinger and Fernandez-Cornejo [1995]).

FIFRA sets out the legal framework for regulating the distribution, sale, and use of pesticides in the United States. FIFRA broadly defines pesticide to include "any substance or mixture of substances intended for preventing, destroying, repelling, or mitigating pests" (Butts et al.[1995]). Before a pesticide can be distributed or sold in interstate or intrastate commerce, it must be registered or licensed with the EPA. The registration is based on data "demonstrating that the pesticide will not cause unreasonable adverse effects on human health or the environment when it is used according to approved label instructions." Much of the data required for registration addresses such topics as application information, the exposure levels of farm workers and consumers, and the potential risks to the environment and wildlife. Each pesticide may have several uses and each use must have its own registration. Each registration defines the crops or sites on which the pesticide can be applied and how it can be used (i.e. method of application, application rates, timing, target pests, environmental precautions such as application near wetlands, water, or endangered species) (Osteen [1994]). There are, however, provisions in FI-

FRA to allow uses without registration such as emergency exemptions (Section 18) and experimental use permits (Section 5). EPA estimates that about 4,000 registrations are currently in effect. FIFRA also authorizes the establishment of the previously discussed Worker Protection Standards which are designed to protect farm workers through education, notification, personal protective equipment and field re-entry intervals. In addition, FIFRA sets out the Special Review process which applies to pesticides that have already been registered but EPA subsequently determines that their continued use could lead to unreasonable adverse effects on man or the environment.

Pesticides are also regulated by various provisions of the Federal Food, Drug and Cosmetic Act (FFDCA). Under the FFDCA, the EPA establishes the maximum allowable level (tolerance) of pesticide residues that can be present on foods sold in interstate commerce and the Food and Drug Administration (FDA) monitors food and feed for pesticide residues. From the late 1950s through 1995, FFDCA treated residue on processed food differently from that on raw or unprocessed commodities. Prior to enactment of the Food Quality Protection Act of 1996, FFDCA's section 408 tolerance decisions considered "risks and benefits by weighing the need for an adequate, wholesome, and economical food supply against the need to protect consumer's health..." (Osteen [1994]). Section 409 tolerance decisions, however, considered risk only, i.e. a reasonable certainty of no harm. A controversial portion of section 409 of the FFDCA was the Delaney (zero risk) clause. This clause prohibited pesticide residues from processed foods (a tolerance is not granted) if the pesticides or inerts were found to induce cancer in humans or animals (Callahan [1994]). This prohibition applied "regardless of whether or to what extent its residue is judged to pose a hazard to human health," (National Research Council [1987, p. 21]) but did not apply to raw commodities. EPA's coordination policy, however, which applied to pesticides used on raw commodities and residues of which are found in processed food or feed products, resulted in revocation of the tolerance for raw commodity (denial of a new tolerance) if the pesticide concentrates in a processed food or feed product. With the passage of the Food Quality Protection Act in 1996, the distinction between raw and processed foods with respect to pesticide residue standards has been eliminated.

In 1996 Congress passed the Food Quality Protection Act (FQPA) which was intended to update and resolve inconsistencies in the two major pesticide statutes: FIFRA and FFDCA. The major components of the FQPA address the issues of setting a single, health base standard (i.e., a reasonable certainty of no harm) for all pesticides in all foods (although benefits can continue to be considered in certain instances when setting standards), providing special protection for infants and children; expediting approval for safer pesticides, regulatory relief for minor use pesticides, requiring periodic reevaluation of pesticide registrations and tolerances and reauthorizing and increasing registrant fees to fund such reevaluations, establishing national uniformity of tolerances unless States petition for an exception, and mandating the distribution of information in grocery stores on the health risks of pesticides and how to avoid such risks (Pesticide and Toxic Chemical News [August 21, 1996] and Environmental Protection Agency [1996]).

Other statutes with the potential to affect pesticide/fertilizer use include the Clean Air Act, Clean Water Act, Safe Drinking Water Act, Coastal Zone Management Act (CZMA), Endangered Species Act, Resource Conservation and Recovery, and the Comprehensive Environmental Response, Compensation, and Liability Act of 1980 (Superfund). The Clean Water Act of 1972 (sec. 319) and the CZMA address non-point sources of pollution, such as that from farm fields. Under the former, States can establish a management plan to address non-point pesticide and fertilizer pollution from agricultural sources. The CZMA creates the non-point pollution control program which increases the affected States' authority to implement control measures on the use of fertilizers and pesticides. The CZMA authorizes federal grants to States which agree to establish coastal management plans to reduce non-source pollution. States, in turn, are encouraged to assist producers in constructing individual nutrient and pesticide management plans (Environmental Protection Agency [1995]).

The Clean Air Act of 1970 sets emission standards for fertilizer and pesticide plants and specifies the type of equipment necessary for compliance. Of particular concern to agriculture is the announced phaseout of methyl bromide production and use in the U.S. by the year 2001, which is an action under the Clean Air Act. This action is based on the determination that methyl bromide is an ozone depleting chemical. Methyl bromide is a wide-spectrum pest control fumigant used on a large number of vegetable crops and storage structures (Osteen [1994]).

The Safe Drinking Water Act sets national standards for levels of contaminants, such as pesticides and fertilizers, found in drinking water from public water supplies. National drinking water standards include the legally-enforceable Maximum Contaminant Levels (MCL). Repeated violations of MCL's may result in the closure of the public water system and require the local jurisdiction to provide alternative sources of drinking water.

The Endangered Species Act (ESA) instructs EPA to take steps to prevent harm to endangered species or its habitat from pesticides. The ESA regulations concerning pesticides have not been finalized, but initial efforts concentrated on identifying vulnerable species and their location, specifying which pesticides "may affect" endangered species, and changing the label application rate if the endangered species are near the application site.

The Resource Conservation and Recovery Act of 1976 specifies how organizations should contain and dispose of toxic substances. Superfund legislation stipulates who pays for existing toxic dump sites and establishes a trust fund for dump site clean-ups.

There are a number of instances in agriculture where a regulatory approach has been used or suggested as a way of controlling for the externality arising from production. Currently, at least 17 states have enacted regulations governing fertilizer and nutrient use by farmers. No state, however, has a comprehensive legal framework for protecting both surface water and groundwater from all agricultural nonpoint source pollution (Ribaudo and Woo [1992]). State controls range from requiring a farmer to obtain a permit to apply nutrients, to banning certain management practices, to restricting chemical use.

An example of State-level chemical use controls is Nebraska's Groundwater Management and Protection Act, designed to reduce nitrate concentrations in groundwater by controlling fertilizer use. This act established broad local control for solving water quality problems by giving extensive management responsibilities to the 23 Natural Resource Districts (NRD) covering the state. The NRDs may act alone or in cooperation with the Nebraska Department of Environmental Quality to develop management plans that may require farmers to adopt specified management practices. Since the passage of the act, at least three NRDs have established ground water control areas. The most important and extensive of these is the Central Platte NRD, where fall and winter applications of nitrogen fertilizer on sandy soils are restricted, and in the most severely affected areas, farmers are required to conduct nitrogen tests and keep records of nitrogen fertilizer applications. The regulations do not, however, require farmers to actually use nitrogen test results or to restrict the amount of fertilizer applied (Williamson [1988]).

IMPLICATIONS OF AGRICULTURAL CHEMICAL REGULATION

The two major statutes, FIFRA and the FFDCA instruct regulators to weigh the benefits of pesticide use against "unreasonable" risks. Wilkinson has succinctly characterized this balancing process: "While the use of the term 'unreasonable risk' implies that some risks will be tolerated under FIFRA, it is clearly expected that the anticipated benefits will outweigh the potential risks when a pesticide is used according to commonly recognized, good agricultural practice" (National Research Council [1993]).

A study of the impact of pesticide regulation on innovation and the market structure in the U.S. pesticide industry (Ollinger and Fernandez-Cornejo [1995]) shows that pesticide regulation in the United States has encouraged the introduction of fewer, yet less toxic, pesticides.[15] The 1972, 1978, and 1988 amendments to FIFRA require that new and existing pesticides meet strict health and environmental standards. Requirements for pesticide registration with the EPA include field testing that can include up to 70 different types of tests that can take several years to complete and cost millions of dollars. They consist of toxicological studies, a two-generation reproduction and teratogenicity study, a mutagenicity study, oncogenicity studies, and chronic feeding studies. The toxicological studies include acute (immediate), subchronic (up to 90 days), and chronic (long term) effects. Other tests are used to evaluate the effects of pesticides on aquatic systems and wildlife, farm worker health, and environmental fate. Recent estimates suggest that research and development of a new chemical pesticide (including the testing indicated above) costs between $50-70 million and takes 11 years (Ollinger and Fernandez-Cornejo [1995]). As a consequence of the regulation requirements, pesticide firms refocused their research away from persistent and toxic pesticides. The number of pesticides with chronic (long term) toxicity dropped by 86 between the 1972-76 and 1987-91 periods and lower toxicity pesticides account now for more than half of the pesticide sales. This related to the reduction in chronic TPUs of applied pesticides discussed in Chapter 2. A 10 percent increase in testing costs is associated a 2.8 percent increase in the proportion of "less toxic" pesticides registered.

---

[15] High toxicity pesticides are those that belong to Class I acute toxicity (indicated on the label), or are chronically toxic to humans, or fish and wildlife. Lower toxicity pesticides are all others (Ollinger and Fernandez-Cornejo [1995]).

Pesticide regulation has also had undesirable consequences. Regulation discouraged new chemical registrations: The number of new pesticides registered by the EPA in 1987-91 was half that of 1972-76 and each 10 percent increase in pesticide regulatory costs caused a 2.7-percent reduction in the number of new pesticides introduced (Ollinger and Fernandez-Cornejo [1995]). The higher regulatory costs contributed to an industry-wide increase in research spending which encouraged some small firms to leave the pesticide industry. Pesticide regulation also encouraged firms to focus their research on pesticides used in larger crop markets such as corn and soybeans abandoning minor crop markets, such as horticultural crops. The decline in new registrations of chemical pesticides suggest that there are market opportunities for biological pesticides (e.g., viruses and pathogenic bacteria) and genetically modified plants. These products are not only environmentally preferable but also less costly to develop and register than chemical pesticides. These new products, however, are only effective against a narrow range of pests (Ollinger and Fernandez-Cornejo [1995]).

Cropper et al. [1992a] examined EPA's Special Review Process for pesticides between 1975 and 1989 to determine whether the decision to cancel or continue the registration of pesticides could be explained by the risk and benefits associated with pesticides. Cropper et al. estimated a trade-off of $35 million in producer benefits per cancer avoided among pesticide applicators. More recently, Cropper et al. [1992b] estimate a trade-off of $72 million per cancer avoided among pesticide applicators and $9 million per cancer avoided among consumers (in 1986 dollars). Abler [1992] argues that these figures are too high because Cropper et al. calculated the benefits at existing prices, not considering the effect of pesticide restrictions on producer prices. Abler argues that producers could even gain from pesticide restrictions if output prices increased enough. On the other hand, higher output prices caused by restrictions on pesticide use have not been empirically documented.

The use of nitrogen fertilizer has been singled out for special consideration because it appears that application of nitrogen fertilizer can be reduced without adversely impacting crop yields. National, regional, and farm level nitrogen mass balances suggest that current nitrogen inputs from all sources usually exceed the nitrogen harvested and removed with crops (Fuglie and Bosch [1995]). For example, if all sources of nitrogen including synthetic fertilizers, manure and that fixed by legumes are accounted for, estimated nitrogen inputs are nearly 1.5 times as great as the nitrogen removed in harvested crops or crop residues (National Research Council [1993]).

Peterson and Frye [1989] have estimated the amount by which the nitrogen in synthetic fertilizers applied to corn acreage (aggregated to the national level) replaced the amount of nitrogen removed in the grain of that year's crop. The nitrogen applied to corn in synthetic fertilizer exceeded that removed in the grain by 50 percent or more for every year since 1968. Similarly, Peterson and Russelle [1991] estimated the amount of nitrogen applied to corn in synthetic fertilizers and that supplied by alfalfa in the Corn Belt. Depending on how much nitrogen was supplied by alfalfa, they estimated that nitrogen applications could be reduced by between 8 and 14 percent for the region as a whole without adversely affecting yields. For states such as Michigan, Minnesota, and Wisconsin which grow relatively more alfalfa than other states, the estimated nitrogen reductions were larger, 20 to 36 percent for Michigan, 13 to 23 percent for Minnesota, and 37 to 66 percent for Wisconsin. Alternatively, Legg et al. [1989] estimated the total nitrogen per acre applied from all sources was, on average, 64 pounds per acre in excess of the nitrogen needed to achieve yield goals. Similar results have been reported elsewhere (National Research Council [1993]). For example, crop production budgets for Nebraska suggest that since the mid-1960s, the amount of nitrogen applied to cropland has exceeded crop requirements by 20 to 60 percent (Olson [1985]).

National and/or regional mass balances are crude generalizations of the real situation in particular field crops. The actual balance between nitrogen applied and that applied for crop growth varies from region to region, from farm to farm, and even from field to field (Economic Research Service [1997]). Nitrogen must be applied in excess of the amount actually harvested in grain and residues because the efficiency of nitrogen uptake by the crop is less than 100 percent and because precise crop needs vary with time and weather. The magnitude of the unaccounted for nitrogen in estimated mass balances, however, indicates the underlying reason for the loss of nitrogen from crop production and illustrates the potential for improvements in nitrogen management.

This being the case, in a study of the effects of the Nebraska Groundwater Management and Protection Act, Bosch et al. [1995] found that for Central Nebraska, farmers whose fertilizer applications were subject to direct control were more likely to adopt fertilizer testing. Moreover, the test results were utilized in making nitrogen fertilizer application decisions (ostensibly resulting in lower application rates) compared to farmers outside of the region. In a companion piece, Fuglie and Bosch [1995] find that when uncertainty about the quantity of soil nitrate is highest, the value of soil nitrogen testing is highest, enabling farmers to reduce nitrogen fertilizer use without affecting crop yield.

In another study focusing on restricting fertilizer use, Quiroga, Fernandez-Cornejo, and Vasavada [1995] use the virtual pricing approach to assess the impact of reducing fertilizer use on agricultural production in Indiana. The study concludes that, based on data covering the 1950-1986 time period, if fertilizer use is restricted, the agricultural sector would adjust by expanding the use of labor, feed, and pesticides. Seed and fuel use are not significantly affected. Profits become negative if fertilizer use is reduced by more than 14 percent.

The most extreme form of regulation is a ban on use, or, more technically, canceling the registration required for use. Probably the most famous instance of banning the use of agricultural inputs are those of the organochlorines class of pesticides. It was determined that many of the original organochlorines dating to the 1940s resulted in severe externalities and presented unacceptable risks to human health (National Research Council [1989]). Most of these compounds have been removed from agricultural use and replaced with less persistent organophosphates, carbamate, and synthetic pyrethroid pesticides. Because the substitutes were generally as effective as the pesticides they replaced, little was lost in terms of agricultural productivity. The environmental benefits were ostensibly substantial although there is no consensus on the precise magnitude of these benefits.

Several studies consider the effect of banning pesticides or fertilizers in general or banning specific pesticides. Knutson et al. [1990], using a mathematical programming framework, estimate that the total elimination of chemical use (pesticides and fertilizers) in U.S. agriculture would increase annual consumer expenditures by $428 per household (in 1986 dollars). They also estimate that food prices during 1995-98 would increase at double digit levels as in the 'food crisis' of the 1970s. Grain and cotton exports would decrease by 10 percent and cultivated acreage would increase by 10 percent with a corresponding effect in erosion. Knutson et al. [1994], using a similar programming approach to their 1990 study, calculate sharp yield losses and cost increases in the production of selected fruits and vegetables after a 50 and 100 percent reduction in pesticide use.

Using a linear programming model of production and marketing of major field crops, Taylor and Frohberg [1977] investigate banning herbicides and insecticides in agricultural production in the Corn Belt. Without herbicides, corn and soybean yields decline by 22 percent, consumers' surplus falls by $3.5 billion, and producers' surplus increases by $1.8 billion. Labor use expands by 7 percent due to the additional cultivation required for corn and soybeans. When insecticides are banned, corn yield declines by 2.9 percent while soybean yield remains virtually unchanged. Consumers' surplus falls by $632 million while producers' surplus increases by $531 million due to banning insecticides. In another study, Taylor et al. [1979] quantify the expected effects of withdrawing all insecticides from cotton production in the United States for the period 1977-1985. With estimated regional yield reductions of 5-28 percent and per acre variable costs increasing $1 to $20 per acre, aggregate annual producer plus consumer surplus loss is estimated to be $775 million dollars. Further, Taylor [1995] examines the economic effect of a complete elimination of pesticides in the production of fruits and vegetables. He concludes that the acreage required for production would increase by 2.5 million acres (44 percent), unit production costs would increase by 75 percent, wholesale prices of fruits and vegetables would increase by 45 percent, returns to producers would decrease by 30 percent, retail prices would increase by 27 percent, and domestic consumption would fall by 11 percent.

These types of studies that examine banning pesticide use have not been without criticism. As the realism of multiproduct firms, nonseparable pest control functions, estimation errors, agricultural policies, farmer risk aversion, and consumer preferences for food safety are added to models to evaluate pesticide restrictions, the transparencies of changes in costs, changes in yield (e.g., the removal of Bt from the market resulting in a yield loss of 100 percent for apples under a no pesticide scenario used by Knutson et al. [1994]), and changes in consumer utility are sometimes lost. Differing assumptions and modeling approaches can give different estimates of welfare losses of pesticide regulations (Swanson and Grube [1986], Taylor [1992], and Smith [1995]).

A study prepared by the National Center for Food and Agricultural Policy (NCFAP) considers the impact of the Delaney clause of the FFDCA. The NCFAP study estimates that the aggregate economic loss for 28 crop-pesticide class combinations, selected from a list of 85 crops and 38 pesticide active ingredients that could be affected by the enforcement of the Delaney Settlement Agreement, is $387 million per year (Gianessi and Anderson [1991]).

Based on a sample of 226 cash grain farms in the Lakes States and Corn Belt regions of the United States, Whittaker et al. [1995] evaluate the impact of restricting pesticide use on farm profitability. Focusing on farms that produce soybeans and corn, a linear programming model is used to identify the maximum attainable profits for a farm if it adopts a best management practice. Using this as the point of reference, the study estimates that limiting pesticide expenditures to no more than $22 per acre[16] had almost no impact on profits among small and medium size farms and had a minimal impact on large farms. They suggest that there are pest management practices currently available that can be adopted without significantly reducing farm profits.

A variety of biologic and economic studies of pesticide use estimating the cost to producers and consumers of banning specific pesticides for specific crops have been conducted by USDA's National Pesticide Impact Assessment Program (NAPIAP). Numerous assessments have been conducted beginning in 1977, but since market and pest conditions change often, the focus is limited to those conducted over the last five years to present. The value of certain herbicides, insecticides, and fungicides used on corn, cotton, cranberries, leafy green vegetables, grain sorghum, and soybeans is analyzed. The results are summarized in Table 4.1.

The value of a pesticide's use to producers and consumers is influenced by a variety of factors, including market and pest conditions, and the availability and characteristics of alternative means of pest control. For example, when market price and pest infestation are high, the value of pesticide use is high as well and producers will protect as much of their crop as possible. The cost of not doing so is high. Conversely, when market price and pest pressure are low, the value of pesticide use is low as well. Table 4.1 generally shows the effects for a single year or an average of several years.

[16] In the sample, the average expenditure per acre on pesticides was $21.39 for small farms, $29.88 for medium farms, and $25.60 for large farms.

Table 4.1. Economic Impact of the Loss of Selected Chemical Uses for Selected Crops. Summary of NAPIAP studies

Crop/Chemical	TYPE (H,I,F)	Producer loss	Consumer loss	Total loss
		------------	$ million	------------
Corn				
Captan	F	1465	na	na
Triazines	H	911	na	na
Atrazine	H	679	na	na
Acetamides	H	76	na	na
Granular	I	58	na	na
Granular + liquid	I	42	na	na
Cotton				
Pyrehroids	I	172	na	na
Desiccants/ defoliants	146	na	na	
Dinitroanilines	H	139	na	na
Seed treatments	F	139	na	na
Substituted ureas	H	(28)	na	na
Cranberries				
Groups of herbicides	H	34	31	65
Groups of fungicides	F	20	19	39
Groups of insecticides	I	11	10	21
Leafy green vegetables[1]				
Lettuce	F	127	160	287
Spinach	F	16	3	19
Collards	F	5	na	na
Turnip greens	F	4	na	na
Kale	F	2	na	na
Mustard greens	F	1	na	na
Grain sorghum[2]				
Atrazine	H	(4)-22	54-87	58-65
Glyphosate	H	(1)-1	5-8	6 - 7
Soybeans				
Carboxin	F	67	na	na

Table 4.1. Economic Impact of the Loss of Selected Chemical Uses for Selected Crops. Summary of NAPIAP studies (continued)

Pendimethalin	H	50	na	na
Metribuzin	H	35	na	na
Bentazon	H	32	na	na
Imazethapyr	I	(77)	na	na
Various fruits, vegetables, and non-food crops				
Methyl bromide[3]	Fumigant			
With imports	775	269	1044	
Without imports	300	782	1081	

Type: H = Herbicide, I = Insecticide, F = Fungicide.
[1] Estimates are for all foliar fungicides.
[2] Under some scenarios, producers who do not use the chemical gain more than users lose.
[3] Excluding quarantine uses for imports. Totals may not sum due to rounding.

The availability, cost, and efficacy of alternative chemicals influence the value of a specific chemical's use. Since many individual pesticides have alternatives of about the same cost and efficacy, the marginal value of any single chemical is relatively low since another chemical is most likely a ready substitute for it. The value of a family of chemicals is likely to be fairly high, however, since the likelihood of suitable substitutes is less. For example, the loss of an individual pyrethroid insecticide would cost cotton producers less than $3 million. The loss of the entire class of pyrethroids, however, would cost producers more than $170 million. (Because atrazine accounts for the bulk of triazine use in corn, its value is close to that of the entire family of triazines.) Some chemicals have few if any substitutes (e.g., captan, a fungicide used to treat corn seed, and methyl bromide, a fumigant used for a variety of crops). The loss of such chemicals would have high costs to producers as well as consumers. USDA estimates that producer and consumer losses of banning agricultural uses of methyl bromide is approximately $1.3-$1.5 billion annually (U.S. Department of Agriculture [July 1993]).

If the possibility of using substitutes is lost, either through a government ban or through pest resistance, then the value of a specific pesticide may increase. This is especially pertinent for government regulation since, as Osteen [1993] points out, regulatory decisions are interdependent. Depending on the order in which pesticides are banned, a chemical on the market may have a high economic value to society simply because its substitutes have previously been banned, even though it may pose a greater health risk than its substitutes. In other words, a chemical's benefit may increase over time not due to an increase in efficacy, but by the elimination of its substitutes from the market.

The value of a pesticide will also vary geographically. Producers in areas of heavy pest infestation will place a higher value on chemical use than those in areas free of the pest. Were a specific chemical not available, production may shift geographically as growers in pest-free areas may expand production to make up for shortfalls from growers exiting the sector in more heavily infested regions. While in aggregate producers generally suffer economic costs from the lost use of a chemical, some producers (nonusers) may gain while others (users) suffer losses.

Other studies of specific pesticides include the evaluation of the welfare costs to cancel the insecticide ethyl-parathion on almonds, plums, and prunes, by Lichtenberg, at al. [1988]. This study calculates the impact on current users and nonusers, domestic and foreign consumers, thus recognizing the redistributional impact of pesticide regulation mentioned earlier. While current nonusers would gain about $0.5 million in producers' surplus, the net welfare impact for the U.S. alone is a loss of about $2 million while world welfare would decline by $2.4 million. Regarding the distributional impact among consumers, Zilberman et al. [1991] argue that because pesticide use contributes to reducing food prices, consumers in general, particularly low-income consumers, benefit economically from pesticide use.

In sum, Government regulation and direct controls attempt to internalize the costs of externalities. The most efficient means of farm chemical use, as defined by the market, is blocked by government intervention. This means that costs will be imposed on input suppliers and crop producers, as discussed above. Greater testing costs from regulation may reduce competitiveness among suppliers as smaller manufacturers exit the market. Producers have fewer chemical options to choose from, especially producers of minor crops. The extent to which regulation increases the cost of production will affect the extent to which producers suffer reduced net cash returns. Subsequently, as production falls, price may rise, with an ambiguous effect on net cash returns. In most of the literature, producers are found to suffer losses. Some of the cited studies, however, did not consider price increases subsequent to production decreases and so represent an upper bound of effects. Increased costs may be passed on to consumers in the form of higher food prices, although since the farm share of retail food prices is relatively low, consumer costs will likely be low. The environment will benefit by the removal of a specific chemical use only if more toxic substitutes are not used to replace it.

NON-REGULATORY APPROACHES

Given the choice (apart from a regulatory approach), producers may change production practices for a variety of reasons, some of which include:

1. *Higher profitability*. A new practice may not only be less polluting but also more profitable. Because of inadequate information or uncertainty about the characteristics of the practice, a profit-maximizing farmer may not have adopted them.
2. *Fear of Future Regulation*. A farmer may adopt less polluting practices as a means of pre-empting or forestalling regulations that would be more costly with which to comply.
3. *Recognition of Local Environmental/Health Benefits*. In some cases the environmental amenities enjoyed by a farmer, or his health, can be impaired by the externalities generated as a result of agricultural production. In such cases, education and technical assistance programs can inform a farmer about the consequences of his or her actions, inducing him or her to adopt less polluting practices.
4. *Stewardship*. General concern about the effects of agricultural pollution on the environment and the on the welfare of others can potentially induce a farmer to adopt less polluting practices. Education programs, through moral suasion, will induce some farmers to adopt less polluting practices.

For the first three cases, it is in the farmer's self-interest to adopt a new practice. For the first case, it is clearly "individually rational" for the farmer to adopt the less polluting practice if it yields higher profits or higher expected profits. Self-interest is also the motivating factor for the second case. Fear of regulation is commonly cited in farmer surveys as a reason for participation in voluntary programs (Nowak and O'Keefe [1992]). One of the driving factors behind the agreement between New York City and upstate farmers for protecting the city's water sources was the City's threat to make use of a powerful land-use law. Several states include regulatory components in their watershed management strategies as a way of speeding adoption of best management practices (BMP) (General Accounting Office [1995]). It is unlikely, however, to be politically acceptable for a USDA program to foster this fear in order to induce voluntary adoption.

In the third case, acting in self-interest depends on the farmer believing that his or her actions will improve environmental quality, and that the opportunities for free-riding are limited (Johansson [1987]). The extent of this "free-rider" problem will depend on the number of farmers contributing to a local pollution problem. If the number of farmers is small, and each farmer's contribution is significant and easily observable, the incentive for each farmer to adopt an alternative production practice may be quite strong. If the number of farmers is large, however, and each farmer's contribution is less clear, free-rider problems are likely to undermine voluntary adoption. This is especially plausible in areas where groundwater is the primary source of drinking water and agriculture pollution is responsible for groundwater contamination (Halstead, Padgitt, and Batie [1990]). For these reasons, farmers are more likely to express concern over groundwater quality beneath their own farm than about surface water quality downstream.

Altruism, rather than self interest, is the motivating factor for the fourth case listed above. Although economists tend to dismiss the reliability of altruism as a motivation for altering behavior, some analysts have argued that altruism can play an important role in altering attitudes towards the environment (Baumol and Oates [1979]). At a time when government support to farmers is being reduced and trade is being liberalized, however, the resulting market pressures make it unlikely that the average farmer will adopt costly or risky alternative production practices aimed at controlling externalities for altruistic reasons (Abler and Shortle [1991]).

Various Government policies address these reasons as they seek to influence the choice of practices a producer will pursue. Taxes and subsidies are common policies designed to influence the profitability of a practice. Voluntary educational policies that provide educational, technical, and financial assistance are another set of policies.

(A) TAXES

Historically, taxes were the first policy instrument proposed to deal with the presence of an externality. Pigou [1932] argued in essence that a farmer when using a polluting agricultural chemical which results in an externality does not bear all of the costs of producing the agricultural commodity. In particular, the marginal private cost of production is less than the marginal social cost of the commodity. This can happen when the property rights are not assigned or transaction costs inhibit negotiation between the farmer and those adversely affected by the use of the fertilizer or pesticide. The difference between the marginal private and social costs is the marginal cost of the externality. When this type of externality occurs, the farmer produces more of the agricultural commodity than is socially optimal. This misallocation of resources (market failure) results in a loss of welfare. A solution to the problem is to induce the farmer to supply the socially optimal amount of the good by imposing a tax such that the private marginal cost is increased to the point where it just equals the marginal social cost of production (Meade [1952]).

The implementation of a tax on the use of fertilizer or pesticides has some advantages over other policy approaches dealing with the externalities issue. Because many farmers are involved, the cost of setting and collecting the tax is lower than it is for, say, monitoring nitrate leaching or testing for pesticide residue. Additionally, a tax provides an incentive for a farmer to reduce the amount of the input used. If the farmer reduces the use of agricultural chemicals, the associated externalities will coincidentally be reduced. Next, the production of crops not using the agricultural chemicals giving rise to the externalities will be beneficially affected. Their cost of production will be relatively lower and consequently their output will expand relative to the chemical intensive crops. Thus, such production practices as organic farming relying on minimal chemical inputs will be promoted.

Finally, a tax is preferred to other approaches to controlling an externality because it provides a continuing incentive to the polluting farmer to cut back on emissions. No matter how low they are, cutting back further will reduce tax payments. This is especially important in a dynamic setting, where a polluting farmer is encouraged to seek new low-cost ways of reducing pollution through, for example, adopting an alternative production practice that relies less on the polluting input.

One of the drawbacks to using a tax to adjust for externalities is to determine precisely what the optimal tax rate should be. This requires knowledge of what are the marginal social cost and the marginal private cost of production. The determination of these is frequently an iterative process (Baumol and Oates [1988]). The net cost of reducing externalities and hence environmental quality improvement can be represented as a schedule which is a direct function of the tax rate. A given tax rate will be associated with a minimum net cost for a particular level of environmental quality (e.g., amount of nitrate leached). It has been shown that the pollution abatement cost will increase at an increasing rate as environmental quality is improved (Pfeiffer and Whittlesey [1978]).

A second drawback of a tax is that, if it is decided not to try to recoup the entire marginal social cost of production, it is necessary to determine what is the optimal level of pollution. Unfortunately, there is no generally accepted objective approach to determining such an optimal level. It is based on the available technology (e.g., the ability of farmers to control nitrate leaching via alternative production practices), the capability of accurately monitoring externalities since in the case of agricultural chemicals they are nonpoint source externalities, and society's preference function.

Table 4.2. State Fertilizer Special Taxes

State	Fertilizer Special Taxes
Alabama	None
Alaska	None
Arizona	$0.25/ton
Arkansas	$1.20/ton
California	0.001/$retail sales
Colorado	$0.75/ton
Connecticut	None
Delaware	$0.10/ton semi-annual greater than 10 pounds
Florida	$0.75/ton+$0.50/ton if contains N, $0.30/ton lime
Georgia	$0.30/ton greater than 10 pounds
Hawaii	None
Idaho	$0.15/ton
Illinois	$0.20/ton
Indiana	$0.35/ton inspection fee
Iowa	$0.75/ton
Kansas	$1.70/ton
Kentucky	$0.50/ton ag fertilizer
Louisiana	$0.75/ton
Maine	$0.12/ton
Maryland	$0.25/ton greater than 10 pounds
Massachusetts	$0.15/ton ($5 minimum)
Michigan	$0.10/ton
	$0.015/percent nitrogen/ton fertilizer
Minnesota	$0.15/ton plus $.10/ton surcharge
Mississippi	$0.25/ton

Table 4.2. State Fertilizer Special Taxes (continued)

Missouri	None
Montana	$0.60/ton
Nebraska	$0.10/ton, $5 minimum
Nevada	$0.25/lb, $50/bulk per pound
New Hampshire	$0.20/ton, $5 minimum
New Jersey	$0.15/ton
New Mexico	$0.35/ton
New York	$0.10/ton
North Carolina	$0.25/ton
North Dakota	$0.20/ton
Ohio	$0.12/ton
Oklahoma	$0.65/ton
Oregon	$0.40/ton
Pennsylvania	$0.13/ton, $25 less than 15 pounds
Rhode Island	$0.15/ton
South Carolina	$0.25/ton
South Dakota	$0.22/ton inspection fee, $0.22 regulatory fund
Tennessee	$0.20/ton
Texas	$50/product less than 5 pounds
Utah	$0.15/ton biannually
Vermont	$0.25/ton, $25 minimum
Virginia	$0.25/ton, $35 minimum
Washington	$0.18/ton, Bulk fertilizers $25 license fee for each distributor
West Virginia	$0.30/ton, $15 inspection fee
Wisconsin	$1/ton
Wyoming	None
Puerto Rico	$0.15/ton/yr

Source: Meister Publishing Company [1996]

A further limitation to using a tax to limit an externality is that if is targeted at an especially vulnerable area, a tax sufficient to be effective in mitigating the externality would likely stimulate "bootlegging" of fertilizer or pesticides from nonaffected areas thereby imposing additional costs on those paying the tax on the input.

Using a tax to correct for the externality raises the issue of precisely where to place the tax. In the case of nitrogen which leaches into the groundwater, for example, should the tax be placed on the offending input or should the tax be placed on just the nitrate leached? It has been shown (Holtermann [1976]) that the placement of the tax will affect net farm income with the tax on the actual leaching resulting in the greatest net farm income but the lowest tax revenue.

Most states tax fertilizer, although most also exempt farmers from such taxes. Table 4.2 reports the applicable tax rates for 1996, all of which are placed directly on fertilizer. Farmers must pay the tax in Iowa, Nebraska, Tennessee, and West Virginia (The Fertilizer Institute [1993]). The tax rates are relatively modest given the price of, say, anhydrous ammonia fertilizer is about $300 per ton (U.S. Department of Agriculture [April 1996]). Thus, the tax will raise the price of fertilizer to a farmer somewhere between zero (in states that have no fertilizer tax) and $0.10 to $0.75 per ton (in the states where a farmer is not exempt from the fertilizer tax).

Only a few States tax pesticides and these are mainly lump sum taxes. The exceptions are California which has a tax rate of $0.22 per dollar of sales, Iowa which has a tax rate of 0.1 percent on retail sales, Michigan which has a 0.75 percent tax rate on non-specialty pesticide sales in Michigan, and Minnesota which has a tax rate of 0.1 percent of annual gross sales in Minnesota.

(B) IMPLICATIONS OF A TAX

What has been the impact on net farm income and the environment of existing fertilizer taxes as currently configured? Unfortunately, there is a dearth of empirical assessments of the effects of existing taxes. The impact, however, is likely very modest at best. A zero to 0.5 percent higher price of fertilizer will have an imperceptible effect on fertilizer use.

A number of simulation studies have addressed the question of the impact of higher agricultural chemical prices on agricultural production and the environment. Most are limited in scope, addressing a limited geographic area or a small category of agricultural chemicals using primarily partial equilibrium models. As noted, a tax is more easily implemented than other policy options and its compliance can be monitored more efficiently. Thus, these simulation studies typically do not include any sort of transaction or compliance costs. The extent of the simulated reduction in agricultural chemical use or loss to the environment depends on (1) the structure of the model being used and (2) the magnitude of the assumed price elasticity of demand for agricultural chemicals.

A variety of studies focus on a fertilizer tax. Garcia and Randall [1994] look solely at the production impacts for U.S. corn and wheat of a 10 percent tax on fertilizer using a translogarithmic cost function. They find such a tax reduces fertilizer use by 9.54 percent for corn and 8.76 percent for wheat. This results in corn production in the United States falling by 3.95 percent and wheat production by 2.16 percent. They do not estimate environmental effects.

Quiroga, Fernandez-Cornejo, and Vasavada [1995], using a different methodology, measure the fertilizer tax increase necessary to reduce fertilizer use by 30 percent. They conclude that a tax of 417 percent would be required.

Pfeiffer and Whittlesey [1978] consider a nitrogen fertilizer tax as the approach to abate water pollution in the Yakima Basin of Washington. Farms in the region are typically small and much of the drainage is subsurface during at least part of its return flow to the Yakima River. The target maximum nitrate concentration in the river during August was 0.30 mg/l. The tax required to meet this objective was determined via a linear programming model to be $0.60 per pound. While the fertilizer tax is relatively large (given an assumed fertilizer price of $0.15 per pound), from the standpoint of ease and cost of administration, the authors conclude that such a policy would not be desirable because it could not be administered efficiently by already existing irrigation districts.

Huang and Lantin [1993] use the concept of nitrogen mass balance (basically the difference between nitrogen supplied to a field and nitrogen demanded by the growing crop) and have as an objective the reduction of excess nitrogen in the soil. Based on a nonlinear mathematical programming model for Iowa field crop production, they suggest that to reduce excess nitrogen to zero (and thus eliminate all nitrate leaching), a tax rate in excess of 200 percent would be needed if crop rotations are used and nearly 500 percent if continuous cropping is used.

Using a linear programming model of cotton production in California, Stevens [1988] finds that a nitrogen fertilizer tax of approximately 100 percent will reduce nitrate loss by about 34 percent while reducing farm profits by about 6.7 percent from the no-tax situation. The assessment, however, does not incorporate risk. It has been shown that risk is an important determinant of farmer behavior (Hazell and Scandizzo [1974], Kramer et al. [1983], and Huang et al. [1993]).

In almost all of these studies, the magnitude of the tax on fertilizer or pesticides required to have an appreciable effect on the use of the input is relatively large. This, in conjunction with the problems of setting and implementing a tax noted previously, offer some insights into why, to date, meaningful taxes on agricultural chemicals to control for the externalities associated with their use have not been adopted.

Effects of a tax, like all policy options, vary across the economy. Given that price elasticities for farm chemicals are low, input producers would be able to pass most of the tax along to farmers. To the extent supply is diminished, prices would be expected to rise. Consumers would pay higher costs, more likely true for meat and dairy products (from higher feed costs) than for other foods. Taxpayers would benefit from enhanced revenues. Effects on the environment are positive only to the extent a tax reduces fertilizer use and in areas where nitrogen causes an environmental problem. Given results indicating a high tax required for a use reduction, it is unlikely that use and losses to the environment would be reduced significantly.

(c) Subsidies

The companion policy option to taxes as a way of correcting externalities is to provide a subsidy or "green payment." As originally formulated, a subsidy was suggested when a positive externality results from the production of an agricultural commodity. As developed by Meade [1952], when production results in additional social benefits, the solution to the market failure (supplying less of the good than is socially optimal) is to subsidize the farmer so that the private marginal cost is reduced to the point where he or she is willing to supply the socially optimal amount of the agricultural commodity.

The rule of law, however, has entered into the deliberations. The normal rule in law is to give the recipient of an externality the entitlement against the generator - the polluting farmers must pay in order to continue their polluting activities. Calabresi and Melamed [1972] point out that when the recipients can more easily avoid damage or when transactions costs for collective action are higher for the polluting farmers than for the recipients then the best option may be to pay the polluting farmers to reduce or discontinue polluting. This is what is used to justify offering a subsidy or "green payment" to a polluting farmer. That is, the subsidy issue has changed from its original theoretical development to encompass payments to a farmer to mitigate an externality. Subsidies now are suggested as a way to reduce the use of an input directly or to minimize the externality associated with the use of the input.

Baumol and Oates [1988] argue that under certainty taxes and subsidies have the same short run allocation impacts but their long run effects may be different. The use of subsidies as a mechanism for pollution abatement may introduce new sources of pollution as farmers see the gain associated with reduction of a specific pollutant. Another reason for asymmetry between a tax and a subsidy may be risk and risk aversion. Just and Zilberman [1979] suggest that when the possibility of polluting is uncertain (e.g., nitrogen applied may or may not actually leach into the groundwater) and a farmer is risk averse, a subsidy is likely to result in a higher likelihood of pollution than a tax but a lower likelihood than a subsidy under risk neutrality.

Under certainty, a tax and a subsidy can attain the efficient allocation of resources for an externality but with different equity considerations (Baumol and Oates [1988]). The primary purpose of a policy designed to address an externality is allocative. The basic issue is the direction of resource allocation to achieve a desired level of environmental quality [Freeman [1972]).

The U.S. Department of Agriculture offers producers financial incentives to adopt production practices to improve farm chemical management. These are discussed later in conjunction with the educational and technical assistance programs with which they are often associated.

Other assessments of subsidies as an option for controlling externalities associated with agricultural production are more academic than program assessments. For example, Baker [1988] looks at the relative impact on farm profitability and pollution abatement of subsidizing the production of low input crops in Long Island, New York. A farmer is assumed to be operating in a competitive market and his production function is known and linear. Based on this unique set of assumptions, a subsidy is the optimal policy in terms of realizing the maximum reduction in groundwater contamination. Another study assessing the relative merits of a subsidy to control agricultural externalities is that by Shortle and Dunn [1986]. Realizing that agricultural activities result in nonpoint source pollution and that existing programs are somewhat ineffective, Shortle and Dunn look at the efficiency of providing incentives to adopt a management strategy that reduces runoff. They conclude in a generic context that such incentives have the potential of abating runoff and that subsidies provide the farmer with the greatest incentive to choose a practice that will maximize the expected net social benefits (compared to taxes, adhering to Government-set standards, and mandatory use of specified tillage).

(D) Implications of Subsidies

The effects of a subsidy on an environmentally benign practice (or reduced use of an environmentally damaging practice) rely on how effective the subsidy is in inducing change, whether it is set too low or high. Assuming it is set to achieve the desired goal, the effects are almost the reverse of a tax. Producers would benefit only to the extent the subsidy exceeded their cost to implement the practice. Otherwise, producers would realize no net gain. Consumers would realize no gain or loss. Taxpayers, however, would realize opportunity costs in terms of foregone expenditures with revenues spent on producer subsidies. The environment would be affected to the extent subsidies encouraged adoption of favorable practices.

One problem for an agency offering a financial assistance program is to determine the optimal mix of management/production practices for addressing a particular externality problem, and the optimal subsidy rates that induce a level of adoption sufficient to achieve some exogenous pollution goal at least cost (Malik and Shoemaker [1993]). For example, analysis of payments made under the Water Quality Incentives Program (discussed below) showed that adoption rates of 8 to 73 percent could be achieved for a $0 payment, suggesting that some producers may be willing to adopt certain practices without an incentive payment because of the profitability of the practice, provided that they are given sufficient information on the practice (Feather and Cooper [1995]).

(E) VOLUNTARY PROGRAMS: EDUCATIONAL, TECHNICAL, AND FINANCIAL ASSISTANCE PROGRAMS

Experience with past and present voluntary programs indicates that altruism or concern over the local environment plays only a very small role in a farmer's decisions to adopt alternative production practices. A basic requirement for altruism and concern for local environment and health to be the motivating factors for change is that a farmer believes there is a problem that needs to be addressed, and that his actions will make a difference. The motivating factors can be to protect the quality of one's own water supply, or to protect the environment in general.

Surveys of producer attitudes and beliefs towards the relationship between their actions and water quality indicate that farmers generally do not perceive that their activities affect the local environment. For example, a survey of producers in eight states with Demonstration Projects found that farmers tend to see water quality problems as less serious close to home, but more serious as distance increases (Nowak and O'Keefe [1992]). Less than 10 percent believed that pollution problems associated with agricultural production were serious in their own county. In addition, 40 percent felt that their own farming practices had no impact on local water quality. Such beliefs can limit the opportunity to promote self-interest as a reason for adopting alternative production practices. Similar results were found by Lichtenberg and Lessley [1992] in Maryland and Halstead et al. [1990] in Iowa and Virginia. The perception that a water quality problem exists decreases as the focus is turned on the local area or the farmer's own cropland. Less concern is expressed in regions with more commercialized agriculture, even though actual water quality problems are prevalent in those regions.

Further, Camboni and Napier [1994] found that, even though farmers in an Ohio watershed had a positive attitude towards the watershed as a resource and towards the environment in general, they did not demonstrate greater use of conservation or chemical management practices. Producers either did not believe that current water quality conditions constituted a meaningful threat to their health, or that their actions had any impacts on water quality. Pease and Bosch [1994] found that few farmers surveyed in part of Virginia perceived runoff or leaching problems on their own farms. Even if a sampled cropland site was shown to have a high potential for water quality contamination, very few farmers considered the site to be a high risk for water quality. A survey of producers in the Scioto River watershed in Ohio found that farmers perceived only a slight threat to families' health from existing groundwater contamination by nutrients, and did not attribute the cause of the contamination to agriculture. Observed changes in production practices were based on efficiency and revenue gains, rather than on protecting water quality (Napier and Sommers [1994]). These studies indicate a role for an educational approach to improving chemical use management.

Even though personal health, altruism and fear of regulation can play roles, profitability is the single most important factor in a farmer's decisions to voluntarily adopt alternative or new management practices or systems, especially over the long term (Logan [1990], Camboni and Napier [1994], Norton [1995], and Magleby, Piper, and Young [1989]). A survey conducted in the Nebraska MSEA found that over 85 percent of producers believed that water quality problems exist (Nebraska MSEA [1995]). Producers expressed a willingness to consider water quality impacts when making production practice adoption decisions. Willingness to adopt alternative practices, however, was most influenced by concerns about yield reductions, profitability, and capital and labor constraints. Farming is a business, and one would expect that maximizing net farm income is the primary objective of the management of the farm. A producer's bottom line is affected by changes in yields or production costs, so new practices or technologies that affect either are intensely scrutinized by a farmer. Those practices requiring minor, inexpensive changes in existing practices (i.e. record keeping, manure crediting, legume crediting, and irrigation scheduling) are adopted more frequently than those involving more expensive changes (nitrate testing and split applications of nitrogen) (Feather and Cooper [1995], and Nowak [1991]).

If a voluntary program is to succeed in achieving an environmental objective, then the alternative production practices that are desirable from an environmental quality standpoint must also be profitable. Some of the practices that help mitigate externalities and protect and enhance environmental quality that have been shown to be more profitable than conventional practices include conservation tillage, nutrient management, irrigation water management, and integrated pest management (Bull and Sandretto [1995] and Ervin [1996]). Economic attractiveness, however, depends on location, soil type, climate, level of management and crop produced (Fox et al. [1991]).

(I) Adoption Process

Farmers in general tend to make production changes slowly. Even the adoption of a relatively simple, highly profitable technology such as hybrid varieties of crops took an average of nine years (Office of Technology Assessment [1990]). The adoption process generally can be viewed as having five stages (Nowak and O'Keefe [1992]). Initially, farmers are unaware of a new practice (Stage 1). They become aware of new practices through various sources, including neighbors, farm publications, mass media, extension agents, chemical dealers, and crop consultants (Stage 2). Farmers then evaluate the practice in terms of their own operation through educational sources such as demonstration projects, talking with agents, and talking with neighbors who have tried the practice (Stage 3). Farmers then test the practice on part of their farm (stage 4). The ability of a practice to be tested on part of the farm enhances its potential for adoption (Office of Technology Assessment [1991] and Nowak and O'Keefe [1995]). Finally, full-scale adoption occurs if the practice is found to be acceptable (Stage 5). It must be emphasized that adoption will not advance beyond Stage 1 unless farmers have a compelling reason to seek change.

The role of education and technical financial assistance is to speed up the adoption process. Education activities are targeted at Stages 1 and 2. Education and technical assistance are aimed at stage 3. Technical assistance and financial assistance (subsidies) are targeted at Stages 4 and 5. Adoption of an alternative production practice generally does not occur as a consequence of any specific assistance program (Missouri MSEA [1995]). Change occurs in small increments over a period of time. The role of information assumes special significance in the case of new or emerging technologies (Saha, Love and Schwart [1994]). Research has found that producers' choices are significantly affected by exposure to information about new technology.

A number of impediments may place limitations on adoption of particular practices for a particular farm, even when proposed practices appear to be profitable in field-plot studies. Financial and organizational characteristics of the whole operation may be impediments to adoption (Office of Technology Assessment [1990] and Nowak [1991]). Fixed-cost expenditures, asset fixity, total farm budget (on-farm and off-farm), and availability of farm labor may place limits on changes a farmer can make. Recommended practices may also be incompatible with current production practices and management objectives and will not be adopted, even though marginal net returns are positive at the field level. Ownership characteristics can also influence adoption of profitable practices. Owner-operators are more likely to have greater flexibility to adopt more efficient practices than nonowner-operators, who must often get approval from the owner before making production practice changes. Recommended practices may tax the managerial skill of the operator (Nowak [1991]). New chemical management practices often require a greater degree of managerial skill than conventional practices. An operator may not be able to apply the production practice without specialized assistance. Farmers typically make production decisions within a short time frame which may discourage investment in measures that increase returns only over the long-run (Office of Technology Assessment [1990] and Tweeten [1995]). Risk-averse producers do not look favorably upon practices that are perceived as being too risky and, in many situations, alternative production practices are more risky because of the timing and managerial aspects. There is a greater chance that something might go wrong and net returns decline (Fox et al. [1991]).

(II) EDUCATION

The goal of educational activities is to provide a farmer information that overcomes farmer's inability or unwillingness to adopt an alternative production practice. Education activities generally take the form of demonstration projects and information campaigns utilizing print and electronic media, newsletters, and meetings. Information campaigns are most often used to raise farmer awareness about particular problems or issues. The demonstration projects provide more direct and detailed information about farming practices and production systems, and how these systems are advantageous to the producer. Education, along with other assistance efforts, has played a role in the acceptance and use of IPM techniques in vegetable and field crops (Vandeman, Fernandez-Cornejo, Jans, and Lin [1994]).

Farm*A*Syst (FAS) is an example of a successful educational program. FAS integrates the resources of all levels of government and the private sector to develop proactive, cooperative programs to address pollution concerns on the farm. Pilot studies in the United States and Canada consistently show high farmer receptivity to FAS with a willingness to take action to voluntarily reduce high risks by changing production practices and facility design (Knox, Jackson, and Nevers [1994], Anderson, Bergsrud, and Ahles [1994]). A cost-benefit analysis was conducted for Farm*A*Syst in Louisiana (Moreau and Strasma [1995]). The program was found to have positive net economic benefits under three different methods of calculation (averting expenditures, avoidance costs, and contingent valuation). Lower bound estimates for the net benefits to the approximately 10,000 farmers who are estimated to participate in the program range between $3.3 and $6.3 million. The success of Farm*A*Syst indicates that farmers will take action when potential economic or health risks are high and are they are made aware of this. The types of problems, however, Farm*A*Syst addresses require relatively straightforward solutions, such as digging up a petroleum storage tank. Farm*A*Syst has yet to be successfully applied to field management issues and decisions.

Education programs have been shown to be useful complements to "non-voluntary" programs for reducing the externalities associated with agricultural production. Providing farmers with information about alternative agronomic practices is likely to improve the effectiveness of non-voluntary programs by helping farmers correctly implement alternative production practices. This conclusion is supported by an examination of nitrogen testing in Nebraska (Bosch, Cook, and Fuglie [1995]). Regulation alone does not insure that farmers will use nitrogen testing properly. While regulation was found to be more effective than education in getting farmers to adopt nitrogen testing, adopters in the boundaries of the Nebraska Demonstration Project were more likely to make proper use of the information provided by the nitrogen test. In other words, farmers in regulated areas adopted the test to fulfill the requirements of the regulation, but did not necessarily make use of the information and apply nitrogen fertilizers more efficiently. Regulation alone did not have an "educational" effect. Even where non-voluntary approaches are used to induce farmers to adopt alternative management practices, education is still required so that farmers use the practices properly.

(III) Technical Assistance

When a farmer becomes aware of alternative production practices that can address an externality problem, the availability of technical assistance can speed up the adoption process by overcoming some of the technological and informational constraints that he or she might face. Technical assistance is the direct, one-on-one contact provided by an assisting agency for the purpose of providing a farmer with the planning and knowledge necessary to implement a particular practice on the individual farm. Requirements for successful implementation vary between individual farms because of resource conditions, operation structure, and owner/operator managerial skill.

The availability of technical assistance is often critical, especially for practices that require greater management such as IPM, nutrient management, and precision farming (Dobbs, Bischoff, Henning, and Pflueger [1995]). Where a practice is seen as being incompatible with current farming operations or goals, technical assistance is a means for adapting the practice to better fit the operation, and for suggesting ways the operation might be adjusted as well.

(IV) Financial Assistance

The role of financial assistance (some form of subsidy) is to defray some or all of the installation and start-up costs associated with less-polluting management practices, with the expectation that farmers will continue to use the practice without annual cost-share payments after the initial installation and startup. Total adoption costs faced by a producer consist of two components - investment costs and adjustment costs. Investment costs include the direct costs of implementing the alternative production practice. Adjustment costs include lost production, increased risk, or increased management costs due to learning how to use the alternative production practice efficiently. Cost-sharing has traditionally been used to help farmers adopt or install a structural practice by covering a portion of installation costs. Financial assistance can be used to acquire additional labor or expertise where a farmer's lack of these is an impediment to adoption. More recently, incentive payments are being offered for broader management goals by covering increased risk of lost production during the period the producer is learning how to implement the alternative production practice.

Cost-sharing has been found to be a valuable tool in getting farmers to adopt new management tools that may enhance private returns even only slightly. Because of inertia, farmers are sometimes reluctant to adopt alternative production practices even though return might be higher. The availability of cost-sharing is a way to draw attention to a new practice. Such was the conclusion of a study on nitrogen testing in four Water Quality Program study areas. Public policies to encourage adoption, such as cost-sharing, might be necessary to achieve significant improvements in practice adoption and environmental quality (Bosch, Fuglie, and Keim [1994]). In the Big Darby HUA project, where there was no immediate water quality crisis, financial assistance helped make nonpoint-source pollution more of a priority for some people (General Accounting Office [1995]). An evaluation of Rural Clean Water Program (RCWP) concluded that the availability of cost-share assistance is the most important factor in obtaining producer participation in voluntary nonpoint source control programs (Magleby, Piper, and Young [1989]).

From a government perspective, the optimal financial assistance rates are those that induce the adoption of desired practices at the least cost. Given the administrative burdens of a large financial assistance program, there is a tendency to specify a uniform subsidy rate across inputs and across resource conditions. Uniform rates, however, invariably introduce production distortions. Farmers' decisions are affected by their economic conditions and the resource base. Since these vary, no single cost-share rate is the "right" one for all producers. Allowing subsidies to vary across inputs and resource conditions would reduce these distortions and result in a more efficient program (Caswell and Shoemaker [1993]).

Another complication arising from variations in production and resource allocation across a project is that different production units may need different sets of practices to achieve the same environmental goal. Reductions in pesticide loadings to receiving water may be achieved through reductions in pesticide applications, changes in timing of applications, or changes to the land surface (i.e. field borders and filter strips). A set of practices that is appropriate for one farm may be inappropriate for another. An efficient assistance program requires that program managers correctly identify the appropriate set of management practices for each farm, and devise the appropriate cost-share program to get farmers to adopt the "correct" practices at least cost (Caswell and Shoemaker [1993]). These considerations point to the need for good site planning (technical assistance) coupled with a flexible cost-share program for making the most efficient use of program dollars.

Table 4.3 Major Practices Installed under the Water Quality Incentives Program, Fiscal Years 1992-1994

Practice	Acres
Conservation Cropping Sequence	121,633
Conservation Tillage	95,980
Crop Residue Use	46,521
Irrigation Water Management	111,207
Nutrient Management	239,205
Pasture and Hayland Management	88,645
Pest Management	179,640
Waste Utilization	84,203

Source: Economic Research Service [1995]

Note that one acre treated in two different years with the same practice is counted as two acres treated.

Table 4.4. Water Quality Incentive Practices and Payments

Practice	Payment ($)	Unit	P[1]	N	S	H
Brush Management	25	acre	N[2]	N	N	N
Conservation Cover	15	acre	M	S	L	L
Conservation Cropping Sequence	5	acre	M	S	M	M
Conservation Tillage	12	acre	M	S	M	L
Contour Farming	12	acre	S	M	L	L
Cover and Green Manure Crop	12	acre	L	L	M	L

Table 4.4. Water Quality Incentive Practices and Payments, (continued)

Critical Area Planting	35	acre	N	N	S	S
Crop Residue Use	5	acre	N	N	N	N
Deferred Grazing	7	acre	M	M	M	S
Filter Strip	15	acre	L	L	M	S
Grasses and Legumes in Rotation	15	acre	M	M	S	L
Grazing Land Mechanical Treatment	8	acre	M	M	M	L
ICM-Nutrient Management	10	acre	N	S	N	S
ICM-Pest Management	10	acre	S	N	N	S
ICM-Waste Utilization	10	acre	M	S	S	S
ICM-Conservation Cover	10	acre	M	S	L	L
ICM-Conservation Cropping Sequence	10	acre	M	S	M	L
Irrigation Water Management	10	acre	M	M	S	S
Livestock Exclusion	7	acre	M	S	M	S
Mulching	10	acre	S	S	S	M
Pasture and Hayland Management	7	acre	S	M	M	L
Pasture and Hayland Planting	15	acre	M	S	M	L
Planned Grazing	2	acre	L	M	S	L
Prescribed Burning	5	acre	S	M	M	L
Proper Grazing Use	3	acre	M	M	S	L
Range Seeding	5	acre	M	S	M	L
Record keeping	0.25	acre	M	M	M	L
Regulating Water in Drainage System	10	acre	M	S	M	L
Row Arrangement	12	acre	M	M	M	L
Strip Cropping-						
Contour	7	acre	S	M	M	M
Field	7	acre	S	M	S	M
Wind	7	acre	M	M	L	L
Toxic Salt Reduction	7	acre	M	M	M	L
Waste Management System	0.01	cu ft	M	S	S	S
Waste Utilization	10-15	acre	M	S	S	S

Table 4.4. Water Quality Incentive Practices and Payments, (continued)

Well Testing	150	plan	M	M	M	L
Wildlife Upland Habitat Management	10	acre	S	S	S	S
Wildlife Wetland Habitat Management	10	acre	S	S	S	S
Windbreak Renovation	25	acre	N	N	N	N

[1] P = Pesticide, N = Nutrient, S = Sediment, H = Habitat

[2] N = Negligible, L = Low, M = Moderate, S = Significant

Source: Higgins [1995] and U.S. Department of Agriculture, Soil Conservation Service [1993a].

Financial assistance is not a substitute for education or technical assistance. Even with the availability of financial assistance, a farmer will not adopt an alternative production practice if he or she is unfamiliar with it. The availability of education, technical assistance and financial assistance are critical for maximizing the impact of alternative management practices (Logan [1990]).

As an example of a voluntary approach to deal with externalities associated with the application of pesticides and fertilizers, the U.S. Department of Agriculture had offered two programs to assist producers adopt practices that protect water quality and relate to chemical use: the Agricultural Conservation Program (ACP) and the Water Quality Incentives Program (WQIP). (Both programs have been superseded by the Environmental Quality Incentives Program (EQIP) of the Federal Agricultural Improvement and Reform Act of 1996, which also provides educational, technical, and financial assistance.) Under the ACP, cost-sharing and technical assistance were available to farmers who voluntarily implemented approved conservation and environmental protection practices on agricultural land and farmsteads. While the bulk of cost-share payments went to erosion control, a growing amount and proportion of cost-sharing assistance was being directed to support implementation of practices that deal with fertilizers and pesticides (Table 4.3).

Integrated Crop Management (ICM), which includes pest scouting, nutrient testing, and other improved management practices, was one of the practices included in the ACP. The goal of ICM in particular is to promote the efficient use of pesticides and fertilizers in an environmentally sound and economical manner. ICM encourages farmers to adopt alternative production practices incorporating integrated pest management and nutrient management practices.

The Water Quality Program (WQP) was established in 1990 to protect the Nation's waters from contamination by agricultural chemicals and waste products. The program provided education, technical, and financial assistance only in targeted areas, unlike the ACP. Cost-share payments under the WQP were mainly provided through the Water Quality Incentives Program (WQIP). A purpose of the WQIP was to achieve non-point source reductions in pollution associated with agricultural production of groundwater and surface supplies through the use of incentive payments and technical assistance to farmers who agree to implement approved, non-structural management practices. Eligible producers could enter into multi-year agreements with the U.S. Department of Agriculture to implement approved production practices on their farm, as part of an overall water quality plan, in return for an incentive payment. The WQIP supports 39 different practices for protecting water quality, of which more than 90 percent are rated moderate to significant for pesticide and nutrient concerns (Table 4.4). The incentive payment was meant to compensate farmers for taking on added costs or risk of trying a different production practice.

(f) Implications of Voluntary Programs

The impacts of ICM on water quality cannot be fully determined by changes in input use. Significant improvements in environmental quality can occur by improving input use efficiency, which may not necessarily imply a reduction in actual chemical use. At the same time, reducing annual fertilizer application rates almost always reduces nitrate loadings which have the potential for leaching into the groundwater and running off into the surface water. An analysis of (only) the first year of ICM using data from a sample of four crops grown in four States indicated some success in reducing nitrogen fertilizer use (Osborn et al. [1994]). Nitrogen fertilizer reductions of 16 to 32 percent per acre on corn, wheat, and cotton were found. A second study using 1990 data from 17 states found similar results (Dicks et al. [1991]). Use of other fertilizers (phosphorus and potassium) were found to be largely unaffected. The reported reductions in nitrogen application rates might not be maintainable. To the extent that the initial year reductions accounted for the availability of soil nitrogen accumulations from previous years, nitrogen application rates in future years might have to be increased somewhat as reserves of nitrogen in the soil are depleted.

Dicks et al. [1991] also examined the health and environmental risks associated with the first year of ICM and concluded that risks from pesticide applications were reduced in some instances, while in others they were increased. An index that accounts for risks to farm workers, consumers, and the environment from pesticide applications indicated that ICM had a generally positive overall impact on reducing risks. The impacts, however, were not uniform. About 40 percent of the farms sampled demonstrated a net increase in the index (implying an adverse environmental impact). A major reason was a change in the mix of agricultural chemicals used. Producers switched to pesticides that can be applied at lower rates but leach more readily or are more toxic. Similar results were reported by Osborn et al. [1994] who used a simple screening procedure and concluded that the potential for pesticide leaching increased on some crops in some regions. These results reinforce the fact that a simple goal of reducing pesticide applications may not provide adequate environmental protection from pesticides. Consideration must be given to potential changes in the toxicity or leaching characteristics of new pesticides that might be used, as well as changes in application rates.

Achievements in the WQIP areas include reductions in nitrogen, phosphorus applications and in pesticides as well (Ribaudo [1994]).

The successful promotion of more efficient pesticide management practices affects producers, consumers, and environmental quality. The extent and magnitude of these impacts depends on how widespread adoption becomes. For the individual producer, adoption of a more efficient management practice increases net farm income, by definition. This can be represented by a downward shift in the marginal cost of production. A farm's contribution to aggregate pesticide leaching will be reduced because of the adoption of an alternative production practice.

If the voluntary program is successful and if many producers adopt profit-enhancing alternative production practices, the aggregate supply curve will increase. The consequences of this on producers are ambiguous if the demand for the commodity is inelastic. As production increases, producer surplus may fall if the demand is inelastic. Producers benefit from lower production costs, and greater output, but price may fall as well. Consumers are likely to realize constant or lower prices. If the demand for the commodity is elastic, producers clearly benefit.

The consequences for environmental quality likewise are not clearly defined. At the intensive margin, alternative production practices aimed at pesticides will reduce chemical runoff, by definition. Higher net returns, however, are an incentive for increased production at the extensive margin. It becomes profitable to bring marginal land into production. If this land is more prone to runoff, or if appropriate management practices are not used, externalities might actually increase.

(G) Other Policy Approaches

There have been a number of calls for the reform of Federal programs that support commodity prices and farm income, citing their environmental harm (Curtis [1993]). The Federal Agricultural Improvement and Reform (FAIR) Act of 1996 accomplished some of these reforms. The FAIR Act will affect chemical use by granting producers greater flexibility in their choices of crops to produce, their use of rotations, and by allowing greater acreage to be planted. Direct effects of the FAIR Act may lead to greater chemical use as more acreage is cultivated, but this will be offset to the extent increased use of crop rotations curb fertilizer and insecticide use.

In the past, the Commodity Credit Corporation (CCC), a quasi-public corporation operated mainly by the Farm Service Agency of the USDA, has supported farm income and commodity prices through a variety of programs at varying times. For farm income, the CCC provided deficiency payments to producers who voluntarily participated in the program based on: (1) the farmer's base acreage (average acreage eligible for payments), (2) the program (average historical) yield per acre, and, (3) the difference between a target price (the price of a commodity established by law) and the often lower market price. Since such assistance encouraged surplus production, which lowered farm price and increased CCC payments, the CCC sometimes required participating producers to set-aside a portion of their land. To support farm price, the CCC provided nonrecourse loans to producers, who would pledge their harvest as collateral. This helped support a floor price near the CCC-set "loan rate," which was announced in terms of dollars per unit of production. If the market price was above the loan rate plus interest charges, it would be profitable for the producer to sell the harvest and repay the CCC. If the market price was less than the loan rate plus interest charges, however, the producer could forfeit the appropriate amount of harvest. The CCC accepted title of the collateral and had no recourse to collect additional payments.

The FAIR Act makes significant changes to these forms of support. While retaining nonrecourse loans, the FAIR Act eliminates deficiency payments and replaces them with set and declining payments available to producers who have recently participated in specific commodity programs. Set-aside requirements are eliminated as are requirements to maintain acreage planted to certain crops (base acreage). Hence, producers now have greater flexibility to make their planting decisions more on market signals than on Government programs.

The effects of CCC programs on chemical use has been studied widely.[17] The majority of research suggests that commodity programs have increased chemical use in agricultural production (Ribaudo and Shoemaker [1995], Shoemaker et al. [1990], Hertel et al. [1990], Ogg [1990], Young and Painter [1990], Goldstein and Young [1987], National Research Council [1989]). The findings, however, were not supported in other studies (Gargiulo [1992] and Heady and Yet [1961]). Farm programs influence agricultural chemical use through several means (Miranowski [1975], Ogg [1990], and Young and Painter [1990]) discussed below.

First, farm programs raised prices and reduced price variations in program crops. Higher prices meant higher values of the marginal product for inputs, and hence motivated additional input use. For risk-averse farmers, the price guarantee provided by target prices or loan rates may have encouraged additional nonland input use.

Second, historical deficiency payments were tied to base acreage and program yields. Base acreage (before 1990) and program yields (before 1985) were computed by means of a 5-year moving average. The need to maintain base acreage for program crops induced inflexibility in cropping patterns (Daberkow and Reichelderfer [1988]). The inflexible cropping patterns created by farm programs were dominated by major program crops (U.S. House of Representatives [1989] and Ek [1989]). To the extent some program crops (mainly corn) use more chemicals than non-program crops (with the exception of fruits and vegetables), the base acreage requirement contributed to increased chemical use. Before 1985, documentation of higher yields contributed directly to increased program payments in future years. Therefore, farmers might have applied additional nonland inputs to increase future program benefits. In 1985, program yields were frozen at 1981-85 levels so the incentive for increasing yields to gain future program benefits was removed.

[17] This discussion draws from Lin et al. [1995].

Third, farm programs may have affected agricultural chemical use through set-aside requirements, though the direction of this effect is uncertain. Because few chemicals were applied to set-aside acres, a reduction in total cropped area may have reduced total agricultural chemical use. Further, cover crops could be planted on set-aside land as green manure and hence reduce fertilizer needs the following year. Also, surplus fixed factors (for example, equipment and management) can substitute for agricultural chemical inputs in the short run (Carlson et al. [1992]). For example, mechanical cultivation can substitute for herbicides and the use of economic thresholds may improve the efficiency of pesticide use. Hence, set-asides had the potential to reduce chemical use.

Farmers, however, may have idled their poorest-quality land. This slippage may have affected per-acre chemical use in two ways: agricultural chemical use (for example, fertilizer) increases with land quality; and set-asides reduce total supply of the program crop, and bids up the price and marginal return of the crop. These imply that per acre chemical applications may have increased on cultivated land. Hence, the potential to reduce chemical use by set-asides may have been offset to some degree.

Short-term effects of the FAIR Act on chemical use stem from greater flexibility provided to producers through elimination of base acreage and set-asides. Elimination of base acreage will facilitate rotations, which could reduce fertilizer and insecticide use. Without the concern of maintaining base acreage to receive Federal deficiency payments, one would expect producers to plant crops for which returns are higher (e.g., corn rather than wheat) where producers have such options. Hence, chemical use will change based on how and where the mix of crops change. For example, if producers plant more corn, a relatively more chemically-intensive crop, rather than wheat, one would expect chemical use to increase. Elimination of set-asides and other acreage reduction programs would encourage increased planted acreage with resulting increased chemical use. Set-asides have been relatively low if not zero, however, for several program crops recently so the increase in planted acreage would not be dramatic. In the aggregate, greater chemical use would be expected if more chemically-intensive crops are planted on existing acreage and greater acreage overall is planted, but increased crop rotations would curb such growth.

In the long run, input use will hinge on the relative marginal productivity and cost among labor, chemicals, and other capital, which the FAIR Act will not alter. While Federal support will fall, market demand is expected to keep commodity prices relatively high by historical standards. Thus market incentives will replace Federal incentives regarding application of chemical inputs. As nominal crop prices decline from 1995/96 peaks, however, and real prices are anticipated to continue to decline, there will be less incentive to apply inputs whose productivity does not increase.

(H) IMPLICATIONS OF OTHER POLICY APPROACHES

The effect of the FAIR Act on crop producers varies by crop. Elimination of base acres will allow producers flexibility to pursue those crops that can be grown in their area with the highest returns. This will tend to level returns among crops. Producers who have few options of crops to grow (e.g., in the Northern Plains), however, may face a loss if returns to their crops are reduced by higher production in other areas of the country.

The impact on input suppliers will stem from greater acreage planted and from greater rotations allowed under the FAIR Act. As slightly greater acreage is brought into production, overall chemical use will likely rise. As crop rotations are used to a greater extent, however, this may curb insecticide and fertilizer use. Hence, aggregate chemical use will depend on the relative effects of these two factors. How much acreage is devoted to more or less chemically-intensive crops will also affect the input industry.

Effects on consumers will be indirect and small since commodity prices account for a small share of retail food prices. Taxpayers will save compared to projected expenditures under the previous farm legislation.

Because Federal policy has been evolving towards market-oriented provisions of the FAIR Act, environmental implications are not drastic: producers have been responding to market signals to an increasing degree since 1985 and idled acreage under annual programs has recently been relatively low. In the short-term, increased chemical use from greater acreage planted would be offset to the extent greater crop rotations reduce chemical use. The Act will not significantly change long run trends in farm or input prices and hence should not alter market incentives for input use. Economically, producers will see a reduction of Federal subsidies and become increasingly reliant on market returns, which will have mixed effects for different types of producers. Consumers will likely not gain or lose significantly from crop provisions of the FAIR Act while taxpayers will gain from lower subsidies to producers.

(I) GRADING STANDARDS[18]

Federal fresh produce grades have been criticized for setting unnecessarily high standards for external appearance (Curtis [1993]). Tight standards allegedly encourage use of chemical pesticides by growers and packers. A proposal is to amend grading standards to reduce emphasis on cosmetic qualities and so reduce reliance on pesticide use.

Grading is the categorization of products according to a set of rules or standards. Grading standards were created to facilitate trade when the buyer cannot examine the commodity offered for sale. They are relevant to the extent they reflect final (consumer) demand. Standards for fresh produce emphasize external attributes, such as cleanness, color, shape, size, and surface defects, as well as internal attributes, such as maturity and decay. Each grade has a set of standards specifying the levels of specific attributes required and the amounts of defects tolerated; higher grades allow fewer defects. Since a grade represents a variety of characteristics, potential transactions can proceed without individual characteristics having to be specified. Grades are set at the Federal and State levels, and some firms have private standards as well. Federal standards may be used as a reference point for buyers who have higher specific-quality criteria. The Agricultural Marketing Service (AMS) is the Federal agency chiefly responsible for helping determine Federal standards and inspecting produce, although some States have their own standards as well.

Reducing the use of pesticides would in some, though not all cases, increase the share of produce with blemishes and other defects (Babcock, Lichtenberg, and Zilberman [1992], Elmer, Ewart, and Brawner [1975], Elmer and Brawner [1988], and Conklin and Thompson [1993]). This will raise the probability that consumers would reject them. Van Ravenswaay [1995] concludes that "consumers place a high value on avoiding damage [to produce] although some value avoidance more than others do. There appears to be a segment of consumers who will accept some damage in return for reduced residue level, but acceptable damage level is low and likely dependent on type of fruit or vegetable." For example, consumers may be willing to accept damage to foods that are not intended to be eaten fresh.

The closer that "perfection" is desired in external appearance, the more intensely will inputs be used to achieve that appearance. High costs are re-

---

[18] Portions of this section rely on Powers and Heifner [1993].

quired to achieve the last degree of damage control and meet high standards. Given high standards, producers may use pesticides to minimize the risk of rejection from the premium grade due to a small amount of damage and subsequently receive a lower price.

(J) Implications of Grading Standards

Lowering grading standards will reduce the risk to producers of rejection of their produce for failure to meet standards. This would remove the risk management aspect of pesticide use and allow lower levels of use. Such a policy change would also help assure market access by those producers who choose not to use high levels of pesticides. Hence, as the grading standards are lowered, supplies of the (revised) highest grade would increase. Prices would be expected to fall not only because of the larger supply, but also because consumers would likely not pay premium prices for more blemished foods. Prices of the lowest quality would rise as fewer supplies would fall into that category.

The distribution of benefits and costs among producers would vary with how cost and produce price change. A producer may earn lower net revenue if the reduction in pesticide cost is less than the reduction in price received. Conversely, a producer may earn higher net revenue if the reduction in pesticide cost is greater than the reduction in price received. This would likely vary regionally depending on pest pressures and may lead to a shifting of areas where certain commodities are produced. The effects of lowering grading standards would most likely temporarily benefit producers who use lower levels of pesticides and would likely lead to reduced per acre pesticide use in the short-term.

Effects are unclear for input suppliers. Lower grading standards could negatively affect pesticide suppliers should they face lower demand for their products. Of those producers who earn lower net revenue per acre, however, some may expand acreage to offset such diminished revenues to cover fixed expenses. The result may increase total pesticide use even as per acre use declines.

Consumers would face mixed effects if standards for external appearance were lowered. They would benefit in the short term since premium produce prices would likely fall as quality declines and supplies increase. Consumers who are particularly concerned about pesticide residues and environmental degradation would also benefit. In the short term, however, some consumers might experience difficulties obtaining produce with appearance attributes they desire.

In the long run, marketers might develop alternative mechanisms to Federal standards, including expanded use of brands and business contracts specifying attributes. They would be able to deliver produce with appearance attributes consumers desire, which might be higher than revised Federal standards. These incentives might eventually be passed back to growers, encouraging pesticide use for limiting appearance defects. Consequently, pesticides use may not change much. Growers and packers who can use brands most effectively probably would gain market share.

Changing grading standards would have no impact on taxpayers. The cost to the Federal Government in setting grading standards and inspecting produce is minimal. Growers and importers must pay a fee for USDA certification of their produce.

Changing grading standards would have temporary, mixed effects, but would not be a long run solution to significantly reducing pesticide use. Such a policy change may eliminate overly stringent standards that do not directly relate to cosmetic quality (e.g., the presence of small numbers of insects in lots destined for local delivery) and help reduce chemical use in some specific cases. It would reduce Government barriers to entry for producers who use fewer chemicals and who also have greater defects. This may encourage less per acre pesticide use in the short term, and reduce the probability of consumer exposure to residues on food. If acreage were to expand, however, total use of pesticides may increase and so increase the environmental and health effects discussed earlier. Consumers would benefit from reduced prices but may not be able to find the quality of produce they desire and for which they are currently willing to pay.

Unless consumer willingness to pay for standardized, blemish-free produce changes, however, marketers will likely specify contract characteristics higher than Federal standards, and pay a premium for such produce. They will be able to deliver produce with appearance attributes consumers desire, which might be higher than revised Federal standards. This change may simply replace tight Federal standards with tight private standards. Thus, in the long run, there may be no change in overall pesticide use in response to lower grading standards. The core of the issue is that many consumers demand standardized, blemish-free produce. To the extent production of such produce requires high pesticide use, market forces will ensure that such inputs are applied to meet that demand regardless of how tight or loose Federal standards are.

(K) From Chemical Use to Risk

The effect of the policies discussed above on chemical use is easier to assess than the resulting health risks. Health risk is composed of both exposure and hazard. The link between use and risk is tenuous in part because while use (at some time) is necessary for exposure, it does not necessarily result in exposure. Further, exposure to an agricultural chemical may not generate high risk if the chemical is relatively harmless. Reducing use and exposure is only one aspect of reducing risk: hazard from chemical use and other sources must be considered as well. Reducing hazard associated with farm chemicals might reduce effectiveness of chemicals, but the source of hazard would shift away from chemicals to other sources; molds and other organisms could contaminate food, reduce food supplies, and raise prices, imposing risk and cost on consumers. Both approaches would raise producers' financial risk from reduced income. To reduce health risk, reductions in exposure and hazard must be balanced.

Trade-offs among affected groups (farm families, farm workers, residents near agricultural production sites, consumers of surface and groundwater, and food consumers) and environmental concerns must also be balanced. For example, less persistent chemicals may need to be highly toxic to kill the target pest quickly. This would imply greater risk from greater exposures for those close to the production enterprise (farm workers and families, residents of producing areas), but less risk to consumers. For policy assessment, the distribution of risk must also be considered.

Evaluation of how specific practices will reduce risk is easiest when only one chemical is in question. For example, analyses involving reduction of nitrogen application is a relatively simple case of reducing over-applications and so reducing exposure. It involved no increase in other chemicals and so risk could be assumed to fall from reduced exposure. The other practices are not as straightforward. A myriad of choices are involved in the production of a variety of commodities in a changing environment.

Risk assessment of changes stemming from practices or policies are extremely complex. Most of the policy instruments discussed above offer means to reduce chemical use, but how risk changes depends on policy implementation. Regulating or banning a chemical use must consider substitutes. If producers were prevented from using one type of chemical and switched to a more toxic choice, then even with reduced use, health risk may actually rise. Likewise, taxing use of one chemical may encourage use of more toxic substitutes. Voluntary approaches may reduce risk if they are successful in reducing use of more toxic chemicals. Purely voluntary approaches may not be sufficient, however, to encourage experimentation and adoption of less hazardous management practices. Reform of Federal farm programs will bring changes in crop acreage and chemical use that are too mixed to say to how exposure and hazard will change. Changing grading standards would likely result in slightly less risk in the short term, though probably little change in the long run.

While assessment of production practices must consider substitute practices, assessment of policies must also consider the distribution of risks. One must consider what or who is exposed to what kind of risk. Is the health of wildlife or an ecosystem of concern, or of humans, and if humans, how is the risk to one group weighed against that of another (e.g., consumers and farm workers)? Further, one must also consider the financial risk to which producers and consumers would be subject without effective means of pest control and yield enhancement. Chemicals are used not only to stop pest damage, but also to avoid pest-induced losses, and to ensure yields. Farmers may feel a need to use chemicals as insurance, for example, in case poor weather will not allow them to perform necessary field operations. Unprofitable practices are not sustainable though they may be environmentally benign. The challenge for Federal decision-makers is to find those policy instruments which reduce health and environmental risks as well as producers' financial risks.

To help minimize such trade-offs, biological pest management methods are among options encouraged. It attempts to avoid chemical use altogether. This is the subject of the next chapter.

REFERENCES

Abler, D.G., "Issues in Pesticide Policy," *Northeastern Journal of Agricultural and Resource Economics*, Vol. 21 (1992), pp. 93-94.

Abler, D.G., and J.S. Shortle, "The Political Economy of Water Quality Protection from Agricultural Chemicals", *Northeastern Journal of Agricultural and Resource Economics*, Vol. 21 (1991), pp. 53-60.

Anderson, J.L., F.G. Bergsrud, and T.M. Ahles, "Evaluation of the Farmstead Assessment System (FARM*A*SYST) in Minnesota," *Clean Water-Clean Environment-21st Century, Volume III: Practices, Systems, & Adoption*, Conference Proceedings, March 5-8, 1995, Kansas City, MO, pp. 9-12.

Anonymous, *Pesticide and Toxic Chemical News*, August 21, 1996.

Aspelin, A.L., *Pesticide Industry Sales and Usage: 1992 and 1993 Market Estimates*, BEAD, Office of Pesticide Programs, U.S. Environmental Protection Agency, Washington, DC, June 1994.

Baker, B., "Incentives and Institutions to Reduce Pesticide Contamination of Groundwater, in D. Fairchild (ed.) *Groundwater Quality and Agricultural Practices*, Lewis Publishers, Inc., Chelsea, MI, 1988.

Babcock, B., E. Lichtenberg, and D. Zilberman, "Impact of Damage Control and Quality of Output: Estimating Pest Control Effectiveness," *American Journal of Agricultural Economics*, Vol. 74 (1992), pp. 163-72.

Baumol, W., and W. Oates, *The Theory of Environmental Policy*, Cambridge University Press, Cambridge, 1988.

Bosch, D., Z. Cook, and K. Fuglie, "Voluntary Versus Mandatory Agricultural Policies to Protect Water Quality: Adoption of Nitrogen Testing in Nebraska," *Review of Agricultural Economics*, Vol. 17 (1995), pp. 13-24.

Bosch, D., K. Fuglie, and R. Keim, *Economic and Environmental Effects of Nitrogen Testing for Fertilizer Management*, Staff Report AGES9413, U.S. Department of Agriculture, Economic Research Service, Washington, DC, April 1994.

Buchanan, J.M., and C. Stubblebine, "Externality," *Economica*, Vol. 42 (1962), pp. 371-384.

Bull, L., and C. Sandretto, "The Economics of Agricultural Tillage Systems," in Soil and Water Conservation Society, *Farming for a Better Environment*, Soil and Water Conservation Society of America, Ankeny, IA, 1995.

Butts, E.R., R.P. Myrick, and T.A. King, "1995 Regulatory File," *Farm Chemicals Handbook, 1995*, Meister Publishing Company, Willoughby, OH, 1995.

Calabresi, G., and A.D. Melamed, "Property Rules, Liability Rules and Inalienability: One View of the Cathedral," *Harvard Law Review*, Vol. 85 (1972), pp. 1089-1128.

Callahan, R. "The ABCs of Industry Regulations," *Farm Chemicals*, pp. I13-I18, September 1994.

Camboni, S.M., and T.L. Napier. 1994. "Socioeconomic and Farm Structure Factors Affecting Frequency of Use of Tillage Systems". Invited paper presented at the Agrarian Prospects III symposium, Prague, Czech Republic, September.

Cantor, P., "Health Effects of Agrichemicals in Groundwater: What Do We Know," *Agricultural Chemicals and Groundwater Protection: Emerging Management and Policy*, Proceedings of a Conference held in St. Paul, MN, October 22-23, 1988, Lewis Publishers, Chelsea, Michigan, pp. 21-32.

Carlson, G., M. Cochran, M. Marra, and D. Zilberman, "Agricultural Resource Economics and the Environment," *Review of Agricultural Economics*, Vol. 14 (1992), pp. 313-326.

Caswell, M., and R. Shoemaker, *Adoption of Pest Management Strategies Under Varying Environmental Conditions*, Technical Bulletin 1827, U.S. Department of Agriculture, Washington, DC, December 1993.

Conklin, N., and G. Thompson, "Product Quality in Organic and Conventional Produce: Is There a Difference?" *Agribusiness*, Vol. 9 (1993), pp. 295-307.

Cooper, J., and R. Keim, "Incentive Payments to Encourage Farmer Adoption of Water Quality Protection Practices," *American Journal of Agricultural Economics*, Vol. 78 (1996), pp. 54-64.

Cropper, M.L., W.N. Evans, S.J. Berardi, M.M. Ducla-Soares, and P.R. Portney, "The Determinants of Pesticide Regulation: A Statistical Analysis of EPA Decision Making," *Journal of Political Economy*, Vol. 100 (1992a), pp. 175-97.

Cropper, M.L., W.N. Evans, S.J. Berardi, and M.M. Ducla-Soares, "Pesticide Regulation and the Rule-making Process," *Northeastern Journal of Agricultural and Resource Economics*, Vol. 21 (1992b), pp. 77-82.

Curtis, J. with L. Mott and T. Kuhnle, *Harvest of Hope: The Potential for Alternative Agriculture to Reduce Pesticide Use*, Natural Resources Defense Council, Washington, 1993.

Daberkow, S., and K. Reichelderfer, "Low-Input Agriculture: Trends, Goals, and Prospects for Input Use," *American Journal of Agricultural Economics*, Vol. 70 (1988), pp. 1159-1116.

Dicks, M.R., P.E. Norris, G.W. Cuperus, J. Jones, and J. Duan, *Analysis of the 1990 Integrated Crop Management Practices*, Circular E-295, Oklahoma Cooperative Extension Service, Oklahoma State University, Norman, OK, 1991.

Dobbs, T.L., J.H. Bischoff, L.D. Henning, and B.W. Pflueger, "Case Study of the Potential Economic and Environmental Effects of the Water Quality Incentives Program and the Integrated Crop Management Program: Preliminary Findings," Paper presented at annual meeting of the Great Plains Economics Committee, Great Plains Agricultural Council, Kansas City, MO, April 1995.

Economic Research Service, *Agricultural Resources and Environmental Indicators*, U.S. Department of Agriculture, Economic Research Service, Washington, DC, July 1997.

Ek, C., *Farm Program Flexibility: An Analysis of Triple Base Option*, Congressional Budget Office, Washington, DC, 1989.

Elmer, H., and O. Brawner, "The Effect of Citrus Thrips and Citrus Red Mite on Navel Orange Fruit Production in California's San Joaquin Valley," *Proceedings of the 1988 International Society of Citriculture*, Los Angeles, CA, 1988

Elmer, H, W. Ewart, and O. Brawner, "Effect of Citrus Thrips Populations on Navel Orange Fruit Yield and Tree Growth in Central Valley," *Citrograph*, Vol. 61 (1975), pp. 9-10, 20-22.

Environmental Protection Agency, *Guidance Specifying Management Measures for Sources of Nonpoint Pollution in Coastal Waterways*, Office of Water, Washington, DC, 1995.

Environmental Protection Agency, *Major Issues in the Food Quality Protection Act of 1996*, Environmental Protection Agency, Prevention, Pesticides, and Toxic Substances, Washington, DC, August 1996.

Feather, P., and J. Cooper, *Voluntary Incentives for Reducing Agricultural Nonpoint Source Water Pollution*, AIB 716, Economic Research Service, U.S. Department of Agriculture, Washington, May 1995.

The Fertilizer Institute, *Summary of State Fertilizer Laws*, The Fertilizer Institute, Washington, DC, 1993.

Fisher, A.C., *Resource and Environmental Economics*, Cambridge University Press, Cambridge, 1981.

Fox, G., A. Weersink, G. Sarwar, S. Duff, and B. Deen, "Comparative Economics of Alternative Agricultural Production Systems: A Review,"

Northeastern Journal of Agricultural and Resource Economics, Vol. 20 (1991), pp. 124-142.

Freeman, A.M., "The Distribution of Environmental Quality," in A. Kneese and B. Bower, eds., *Environmental Quality Analysis: Theory and Method in the Social Sciences*, The Johns Hopkins University Press, Baltimore, 1972.

Fuglie, K., and D. Bosch, "Economic and Environmental Implications of Soil Nitrogen Testing: A Switching Regression Analysis," *American Journal of Agricultural Economics*, Vol. 77 (1995), pp. 891-900.

Garcia, R., and A. Randall, "A Cost Function Analysis to Estimate the Effects of Fertilizer Policy on the Supply of Wheat and Corn," *Review of Agricultural Economics*, Vol. 16 (1994), pp. 215-230.

Gargiulo, C., *Demand for Insecticides in Corn: Effects of Rotations and Government Programs*, Ph.D. dissertation, Department of Agricultural and Resource Economics, North Carolina State University, Raleigh, NC, 1992.

Gianessi, L. P., and J.E. Anderson, *Pesticide Use in U.S. Crop Production: A National Summary*, National Center for Food and Agricultural Policy, Washington, DC, February 1995.

Gianessi, L.P., and J.A. Anderson, *Potential Impacts of Delaney Clause Implementation on U.S. Agriculture*, National Center for Food and Agricultural Policy, Washington, DC, June 1995.

Goldstein, W., and D. Young, "An Agronomic and Economic Comparison of a Conventional and a Low-Input Cropping System in the Palouse," *American Journal of Alternative Agriculture*, Vol. 69 (1987), pp. 51-56.

Halstead, J.M., S. Padgitt, and S. Batie. "Groundwater Contamination for Agricultural Sources: Implications for Voluntary Policy Adherence from Iowa and Virginia Farmers' Attitudes," *American Journal of Alternative Agriculture*, Vol. 5 (1990), pp. 126-133.

Hazell, P., and P. Scandizzo, "Competitive Demand Structures Under Risk in Agricultural Linear Programming Models," *American Journal of Agricultural Economics*, Vol. 56 (1974), pp. 235-244.

Heady, E., and M. Yet, "National and Regional Demand for Fertilizer," *Journal of Farm Economics*, Vol. 41 (1961), pp. 332-348.

Hertel, T., M. Tsigas, and P. Preckel, *An Economic Assessment of the Freeze on Program Yields*, Staff Report No. 9066, U.S. Department of Agriculture, Economic Research Service, Washington, DC, December 1990.

Higgins, E.M., *The Water Quality Incentives Program: The Unfulfilled Promise*, Center for Rural Affairs, Walthill, NE, 1995.

Huang, W., L. Hansen, and N.D. Uri, "Nitrogen Fertilizer Timing: A Decision Theoretic Approach for United States Cotton," *Oxford Agrarian Studies*, Vol. 21 (1993), pp. 41-58.

Huang, W., and R. Lantin, "A Comparison of Farmers' Compliance Costs to Reduce Excess Nitrogen Fertilizer Use under Alternative Policy Options," *Review of Agricultural Economics*, Vol. 15 (1993), pp. 51-62.

Johansson, P. *The Economic Theory and Measurement of Environmental Benefits*, Cambridge University Press, Cambridge, 1987.

Johnson, C.J., P.A. Bonrud, T.L. Dosch, A.W. Kilness, K.A. Senger, D.C. Busch, and M.R. Meyer, "Fatal Outcome of Methemoglobinemia in an Infant," *Journal of the American Medical Association*, Vol. 257 (1987), pp. 2796-2797.

Just, R., and D. Zilberman, "Asymmetry of Taxes and Subsidies in Regulating a Stochastic Mishap," *Quarterly Journal of Economics*, Vol. 153 (1979), pp. 139-148.

Knox, D., G. Jackson, and E. Nevers, *Farm*A*Syst Progress Report 1991-1994*, University of Wisconsin Cooperative Extension, Madison, WI, 1995.

Knutson, R.K., C.R. Taylor, J.B. Penson, and E.G. Smith, *The Economic Impacts of Reduced Chemical Use*, R. Knutson and Associates, College Station, TX, 1990.

Knutson, R.D., C. Hall, E.G. Smith, C. Cotner, and J.W. Miller, "Yield and Cost Impacts of Reduced Pesticide Use on Fruits and Vegetables," *Choices*, First Quarter (1994), pp. 15-18.

Kramer, R., W. McSweeny, and R. Stravos, "Soil Conservation with Uncertain Revenue and Input Supplies," *American Journal of Agricultural Economics*, Vol. 65 (1983), pp. 694-702.

Kross, B.C., G.R. Hallberg, R. Bruner, K. Cherryholmes, and K.J. Johnson, "The Nitrate Contamination of Private Well Water in Iowa," *American Journal of Public Health*, Vol. 83 (1993), pp. 270-272.

Legg, T., J. Fletcher, and K. Easter, "Nitrogen Budgets and Economic Efficiency: A Case Study of Southeastern Minnesota," *Journal of Production Agriculture*, Vol. 2 (1989), pp. 110-116.

Lichtenberg, E., and B.V. Lessley, "Water Quality, Cost-sharing, and Technical Assistance: Perceptions of Maryland Farmers," *Journal of Soil and Water Conservation*, Vol. 47 (1992), pp. 260-263.

Lichtenberg, E., D.D. Parker, and D. Zilberman, "Marginal Analysis of Welfare Costs of Environmental Policies: The Case of Pesticide Regulation," *American Journal of Agricultural Economics*, Vol. 70 (1988), pp. 867-74.

Lin, B., M. Padgitt, L. Bull, H. Delvo, D. Shank, and H. Taylor, *Pesticide and Fertilizer Use and Trends in U.S. Agriculture*, Agricultural Economic Report No. 717, U.S. Department of Agriculture, Economic Research Service, Washington, DC, May 1995.

Logan, T.J., "Agricultural Best Management Practices and Groundwater Protection," *Journal of Soil and Water Conservation*, Vol. 45 (1990), pp. 201-206.

Magleby, R.S., S. Piper, and C.E. Young, "Economic Insights on Nonpoint Pollution Control from the Rural Clean Water Program," Proceedings from the National Nonpoint Source Conference, 1989, pp. 63-69.

Malik, A.S., and R.A. Shoemaker, *Optimal Cost-Sharing Programs To Reduce Agricultural Pollution*, Technical Bulletin 1820, U.S. Department of Agriculture, Economic Research Service, Washington, DC, June 1993.

Meade, J., "External Economies and Diseconomies in a Competitive System," *Economic Journal*, Vol. 62 (1952), pp. 54-67.

Miranowski, J., *The Demand for Agricultural Crop Chemicals under Alternative Farm Program and Pollution Control Solutions*, Ph.D. dissertation, Harvard University, 1975.

Missouri Management Systems Evaluation Area, *A Farming Systems Water Quality Report*, Research and Education Report, Missouri MSEA, 1995.

Moreau, R., and J. Strasma, *Measuring the Benefits and Costs of Voluntary Pollution Prevention Programs in the Agricultural Sector Under Three Alternative Concepts--Averting Expenditures, Willingness-to-Pay, and Avoidance Costs: A Benefit-Cost Analysis of the Farm Assessment System (Farm*A*Syst) as Implemented in Nine Parishes in Louisiana*, Unpublished paper, University of Wisconsin, Madison, WI, 1995.

Napier, T., and D.G. Sommers, "Correlates of Plant Nutrient Use Among Ohio Farmers: Implications for Water Quality Initiatives," *Journal of Rural Studies*, Vol. 10 (1994), pp. 159-171.

National Research Council, *Regulating Pesticides in Food: The Delaney Paradox*, National Academy Press, Washington, DC, 1987.

National Research Council, *Alternative Agriculture*, National Academy Press, Washington, 1989.

National Research Council, *Pesticides in the Diets of Infants and Children*, National Academy Press, Washington, DC, 1993.

National Research Council, *Soil and Water Quality*, National Academy Press, Washington, DC, 1993.

Nebraska Management Systems Evaluation Area, *A Five-Year Summary of Project Accomplishments*, Lincoln, NE, April 1995.

Nielsen, E., and L. Lee, *The Magnitude and Costs of Groundwater Contamination from Agricultural Chemicals: A National Perspective*, AER-576, Economic Research Service, U.S. Department of Agriculture, Washington, 1987.

Norton, N, *Economic Analysis of Factors Affecting the Adoption of Nonpoint-source Pollution-reducing Technologies*, Ph.D. Dissertation, University of West Virginia, 1995.

Nowak, P.J., "Why Farmers Adopt Production Technology," in *Crop Residue Management for Conservation*, Proceedings of a National Conference, Soil and Water Conservation Society, Ankeny, IA, 1991.

Nowak, P.J., and G.J. O'Keefe, "Evaluation of Producer Involvement in the United States Department of Agriculture 1990 Water Quality Demonstration Projects," Baseline Report submitted to the U.S. Department of Agriculture, Washington, DC, November 1992.

Nowak, P.J., and G. O'Keefe. 1995. "Farmers and Water Quality: Local Answers to Local Issues," Draft Report submitted to the U.S. Department of Agriculture, Washington, DC, September 1995.

Ollinger, M., and J. Fernandez-Cornejo, *Regulation, Innovation, and Market Structure in the U.S. Pesticide Industry*, Agricultural Economic Report 719, U.S. Department of Agriculture, Economic Research Service, June 1995.

Olson, R., "Nitrogen Problems," in *Plant Nutrient Use and the Environment*, Fertilizer Institute, Washington, DC, 1985, pp. 115-138.

Ogg, C., "Farm Price Distortions, Chemical use, and the Environment," *Journal of Soil and Water Conservation*, January-February (1990), pp. 45-47.

Osborn, C., D. Hellerstein, C. Rendleman, M. Ribaudo, and R. Keim, *A Preliminary Assessment of the Integrated Crop Management Practice*, Staff Report AGES 9402, U.S. Department of Agriculture, Economic Research Service, Washington, DC, February 1994.

Osteen, C., "Pesticide Use Trends and Issues in the United States," in *The Pesticide Question: Environment, Economics, and Ethics*, D. Pimentel and H. Lehman, eds., Chapman Hall, Ltd., London, 1993.

Osteen, C., "Pesticide Regulation Issues: Living with the Delaney Clause," *Journal of Agricultural and Applied Economics*, Vol. 26 (1994), pp. 6-74.

Pease, J., and D. Bosch, "Relationships Among Farm Operators' Water Quality Opinions, Fertilization Practices, and Cropland Potential to Pollute in Two Regions of Virginia," *Journal of Soil and Water Conservation*, Vol. 49 (1994), pp. 477-483.

Peterson, G., and W. Frye, "Fertilizer Nitrogen Management," in *Nitrogen Management and Groundwater Protection*, R. Follett (ed.), Elsevier Scientific Publishers, Amsterdam, 1989, pp. 183-220.

Peterson, G., and M. Russelle, "Alfalfa and the Nitrogen Cycle in the Corn Belt," *Journal of Soil and Water Conservation*, Vol. 46 (1991), pp. 229-235.

Pfeiffer, G., and N. Whittlesey, "Controlling Nonpoint Externalities with Input Restrictions in an Irrigated River Basin," *Water Resources Research*, Vol. 14 (1978), pp. 1387-1403.

Pigou, A.C., *The Economics of Welfare*, 4th edition, Macmillan Publishing Company, London, 1932.

Powers, N., and R. Heifner, *Federal Grade Standards for Fresh Produce--Linkages to Pesticide Use*, Agriculture Information Bulletin No. 675., U.S. Department of Agriculture, Economic Research Service, Washington, DC, August 1993.

Quiroga, R., J. Fernandez-Cornejo, and U. Vasavada, "The Economic Consequences of Reduced Fertilizer Use: A Virtual Pricing Approach," *Applied Economics*, Vol. 27 (1995), pp. 211-217.

Ribaudo, M., and R. Shoemaker, "The Effect of Feedgrain Program Participation on Chemical Use," *American Journal of Agricultural Economics*, Vol. 87 (1995), pp. 211-220.

Ribaudo, M., and D. Woo, "Summary of State Water Quality Laws Affecting Agriculture," *Agricultural Resources: Cropland, Water, and Conservation*, Economic Research Service, U.S. Department of Agriculture, Washington, September 1992.

Saha, A., H.A. Love, and R. Schwart, "Adoption of Emerging Technologies Under Output Uncertainty," *American Journal of Agricultural Economics*, Vol. 76 (1994), pp. 836-846.

Shoemaker, R., M. Anderson, and J. Hrubovcak, *U.S. Farm Programs and Agricultural Resources*, Agriculture Information Bulletin No. 614, U.S. Department of Agriculture, Economic Research Service, Washington, DC, September 1990.

Shortle, J., and J. Dunn, "The Relative Efficiency of Agricultural Source Water Pollution Control Policies," *American Journal of Agricultural Economics*, Vol. 68 (1986), pp. 668-677.

Skinner, J., "Can We Get More for Our Tax Dollars," *Choices*, Third Quarter (1990), pp. 10-13.

Smith, K.R., "Science and Social Advocacy: A Dilemma for Policy Analysts," *Choices*, First Quarter (1994), pp. 19-24.

Stevens, B.K., "Fiscal Implications of Effluent Charges and Input Taxes," *Journal of Environmental Economics and Management*, Vol. 15 (1988), pp. 285-296.

Swanson, E.R., and A.H. Grube, "Economic Impact of Trifluralin on Soybeans: A Comparison of Selected Estimation Models," *North Central Journal of Agricultural Economics*, Vol. 16 (9186), pp. 769-775.

Taylor, C.R., *Agricultural Sector Models for the United States: Description and Selected Policy Applications*, Iowa State University Press, Ames, IA,, 1992

Taylor, C.R., R.D. Lacewell, and H. Talpaz, "Use of Extraneous Information with an Econometric Model to Evaluate Impacts of Pesticide Withdrawals," *Western Journal of Agricultural Economics*, Vol. 4 (1979), pp. 1-8.

Taylor, C.R., *Economic Impacts and Environmental and Food Safety Trade-offs of Pesticide Use Reduction on Fruits and Vegetables*, Department of Agricultural Economics, Auburn University, June 1995.

Taylor, C.R., and K. Frohberg, "The Welfare Effects of Erosion Controls, Banning Pesticides, and Limiting Fertilizer Application in the Corn Belt," *American Journal of Agricultural Economics*, Vol. 59 (1977), pp. 25-36.

Tweeten, L., "The Structure of Agriculture: Implications for Soil and Water Conservation," *Journal of Soil and Water Conservation*, Vol. 50 (1995), pp. 347-351.

U.S. Congress, Office of Technology Assessment, *Beneath the Bottom Line: Agricultural Approaches to Reduce Agrichemical Contamination of Groundwater*, OTA-F-418, U.S. Government Printing Office, Washington, DC, November 1990.

U.S. Department of Agriculture, *Agricultural Chemical Usage, 1992 Field Crops Summary*, National Agricultural Statistics Service, Washington, 1993.

U.S. Department of Agriculture, *Agricultural Outlook*, Economic Research Service, Washington, DC, July 1993.

U.S. Department of Agriculture, *Agricultural Chemical Usage, 1993 Field Crops Summary*, National Agricultural Statistics Service, Washington, 1994.

U.S. Department of Agriculture, *Agricultural Chemical Usage, 1995 Field Crops Summary*, National Agricultural Statistics Service, Washington, 1996.

U.S. Department of Agriculture, *Agricultural Prices*, Statistical Reporting Service, Washington, DC, April 1996.

U.S. Department of Agriculture, Soil Conservation Service, *Field Office Technical Guide: Conservation Practices Physical Effects*, U.S. Department of Agriculture, Washington, DC, 1993

U.S. General Accounting Office, *Information on and Characteristics of Selected Watershed Projects*, GAO/RCED-95-218, U.S. Government Printing Office, Washington, DC, June 1995.

U.S. House of Representatives, *Low-Input Farming Systems: Benefits and Barriers*, 74th Report by the Committee on Government Operations, U.S. Government Printing Office, Washington, DC, 1989.

Vandeman, A., J. Fernandez-Cornejo, S. Jans, and B-H. Lin, *Adoption of Integrated Pest Management in U.S. Agriculture*, AIB 707, U.S. Department of Agriculture, Economic Research Service, Washington, DC, September 1994.

Van Ravenswaay, E., *Public Perceptions of Agrichemicals*, Council for Agricultural Science and Technology, Task Force Report No. 123, Washington, DC, January, 1995.

Whittaker, G., B.H. Lin, and U. Vasavada, ""Restricting Pesticide Use: The Impact on Profitability by Farm Size," *Journal of Agriculture and Applied Economics*, Vol. 27 (1995), pp. 352-362.

Williamson, D., "Implementation of the Nebraska Nitrate Control Legislation," *Nonpoint Pollution: 1988 - Policy, Economy, Management and Appropriate Technology*, American Water Resources Association, November 1988, pp. 133-39.

Young, D., and K. Painter, "Farm Program Impacts on Incentives for Green Manure Rotations," *American Journal of Alternative Agriculture*, Vol. 5 (1990), pp. 99-105.

Zilberman, D., A. Schmitz, G. Casterline, and J.B. Siebert, "The Economics of Pesticide Use and Regulation," *Science*, Vol. 253 (1991), pp. 518-542.

CHAPTER 5

GOVERNMENT POLICY AND THE DEVELOPMENT AND USE OF BIOPESTICIDES

INTRODUCTION

The continual evolution and adoption of new production practices including relatively pesticide intensive farming has led to a sustained increase in output and coincident benefits to the American consumer in a variety of ways including the price of food. The conventional pesticides used today were new and uncertain in a previous periods. In an analogous way, biopesticides and other forms of biological control being developed today will be conventional pesticides in the future. The growth of biopesticide use is an integral part of the technological revolution in agriculture that has generated major changes in production techniques, shifts in input use, and growth in output and productivity (Carlson and Castle [1972]). Predicting the growth in biopesticide use, however, is difficult due to recent changes in Federal laws affecting the farm sector (a major consumer of biopesticides) and regulating the registration and use of pesticides. Additionally, accurately forecasting the changes in the price of biopesticides relative to the price of conventional pesticides complicate the prediction problem. These issues will be discussed below.

The market for biologically based pest controls is small but fast growing. The market value of biologically based products - natural enemies, pheromones, and microbial pesticides - sold in the United States during the early 1990s was estimated at $95-147 million, 1.3 to 2.4 percent of the total market for pest control products (Office of Technology Assessment [1995]). At least 30 commercial firms produce natural enemies. Even though the current market for biological products is growing and large pest control companies are beginning to participate, the market is still so small that biologicals are unlikely to replace conventional chemical pesticides in the foreseeable future unless major research and development activities are started (Ridgeway et al. [1994]).

Biological pest management includes the use of pheromones, plant regulators, and microbial organisms such as *Bacillus thuringiensis* (Bt), as well as pest predators, parasites, and other beneficial organisms. The U.S. Environmental Protection Agency currently regulates biochemicals and microbial organisms and classifies them as biorational pesticides.

(A) MICROBIAL PESTICIDES AND PHEROMONES

Biorational pesticides have differed significantly from conventional pesticides in that they have generally managed rather than eliminated pests, have a delayed impact, and have been more selective (Ollinger and Fernandez-Cornejo [1995])). Thus, for example, microbial pesticides have not been successful as herbicides because target weeds are replaced by other weeds not affected by the microbial pesticide.

Among the most successful microbials has been Bt, which kills insects by lethal infection. Growers have dramatically increased their use of Bt during the 1990s, especially under biointensive and resistance-management programs, because of its environmental safety, improved performance, selectivity, and activity on insects that are resistant to conventional pesticides. It is one of the most important insect management tools in certified organic production. Bt was used in more than one percent of the acreage of twelve fruit crops in 1995, up from 5 crops in 1991 (Table 5.1). Between 12 and 23 percent of the apple, plum, nectarine, and blackberry acreage received Bt applications in 1995, and it was applied on over half of the raspberry acreage. Among vegetable crops, the acreage treated with Bt increased for 13 of the 20 crops surveyed between 1992 and 1994, and was used on about half or more of the cabbage, celery, and eggplant acreage. Bt has been used on only a couple of field crops. Corn acreage treated was steady at one percent in 1994 and 1995, while treated cotton acreage increased from 5 percent in 1992 to 9 percent in 1994 and 1995.

New Bt strains with activity on insects not previously found to be susceptible to Bt have been discovered in recent years. Current research is devoted to improving the delivery of Bt to pests and to increasing the residual activity and efficacy of Bt.

Table 5.1. Agricultural Applications of Bacillus thuringiensis (Bt), Selected Crops in Surveyed States, 1991-1995

Crop 1	Planted acres	Area receiving application				
		1991	1992	1993	1994	1995
	1000 acres	Percent of acres				
Field crops:						
Corn	64,105	*	*	*	1	1
cotton (upland)	11,650	*	5	8	9	9
Fruit:						
Grapes	796	*	**	2	**	6
Oranges	760	2	**	7	**	3
Apples	345	3	**	13	**	12
Peaches	144	*	**	3	**	5
Prunes	94	*	**	*	**	9
Pears	68	*	**	1	**	2
Sweet cherries	47	*	**	8	**	9
Plums	44	*	**	**	*	14
Nectarines	36	*	**	10	**	22
Blueberries	30	11	**	8	**	5
Raspberries	11	49	**	45	**	52
Blackberries	4	18	**	*	**	23
Vegetables:						
Tomatoes, processed	323	**	6	**	5	**
Lettuce	191	**	18	**	20	**
Sweet corn	164	**	3	**	3	**
Onion	128	**	*	**	1	**
Broccoli	111	**	7	**	14	**
Tomatoes, fresh	104	**	31	**	39	**
Cantaloupe	98	**	32	**	8	**

Table 5.1. Agricultural Applications of Bacillus thuringiensis (Bt), Selected Crops in Surveyed States, 1991-1995, (continued)

Snap beans	71	**	20	**	29	**
Cabbage	70	**	48	**	64	**
Bell peppers	61	**	35	**	37	**
Cauliflower	54	**	12	**	20	**
Cucumbers	51	**	19	**	22	**
Strawberries	46	**	24	**	33	**
Celery	36	**	51	**	61	**
Honey dew	26	**	28	**	10	**
Spinach	10	**	13	**	21	**
Eggplant	4	**	13	**	48	**

* Applied on less than 0.5 percent of the acres.

** Not a survey year for that commodity.

Source: U.S. Department of Agriculture, Economic Research Service, Chemical Use Survey, 1991-1995 (U.S. Department of Agriculture [various])

Pheromones are used to monitor populations of crop pests and to disrupt mating in organic systems and some IPM programs. Pheromones were used on 37 percent of fruit acreage to monitor and control pests and on 7 percent of vegetable acreage to control pests (Table 5.2).

(B) BENEFICIAL ORGANISMS

Natural enemies of crop pests or beneficials may be imported, conserved, or augmented. Many crop pests are not native to this country, and the U.S. Department of Agriculture issues permits for the natural enemies of these pests to be imported from their country of origin. Natural enemy importation and establishment, also called classical biological control, has been undertaken primarily in university, State, and Federal projects. Twenty eight states operate biocontrol programs and most have cooperative efforts with U.S. Department of Agriculture agencies (Office of Technology Assessment [1995]). Some crop pests, such as the woolly apple aphid in the Pacific Northwest, have been largely controlled with this method.

Natural enemies may also be conserved by ensuring that their needs for alternate hosts, adult food resources, overwintering habitats, a constant food supply, and other ecological requirements are met and by preventing damage from pesticide applications and other cropping practices (Landis and Orr [1996]). Over half the certified organic vegetable growers in 1994 were providing habitat for beneficials.

Table 5.2. Use of Selected Biological Pest Management Practices on Fruit and Vegetable Crops, Major Producing States, 1990s.

Crop	Planted acres	Beneficial Insects	Pheromone Traps	Resistant Varieties
	1000 acres	Percent of acres		
Fruit:				
Grapes	730	18	14	31
Oranges	613	22	28	21
Apples	381	2	66	16
All fruits	3251	19	37	22
Vegetables:				
Sweet corn	640	na	17	na
Tomatoes	357	5	6	na
Lettuce	259	3	1	na
All vegetables	2914	3	7	na

na: denotes not available.
Source: U.S. Department of Agriculture, Economic Research Service, Chemical Use Survey, 1991-1995 (U.S. Department of Agriculture [various])

A small but increasing number of companies are supplying natural enemies of insects, weeds, and other pests to farmers. For greenhouse and agricultural crop production, most natural enemies being sold - such as beneficial insects, predatory mites, parasitic nematodes, and insect egg parasites - are used for managing pest mites, caterpillars, citrus weevils, and other insect and arthropod pests. A number, however, of natural enemies - musk thistle defoliating weevils, for example - are being sold for managing weeds on rangeland and uncultivated pastures (Poritz [1996]).

The California Environmental Protection Agency has published a list of commercial suppliers of natural enemies in North American since 1979, and the number has increased steadily. In 1994, 132 companies were listed, mostly in the United States, offering over 120 different organisms for sale (Hunter [1994]).

(C) HOST PLANT RESISTANCE

Corn and soybean breeding for genetic resistance to insects, disease, and other pests has been the research and development focus of major seed companies for many decades (Edwards and Ford [1992]). U.S. soybean acreage, for example, receives virtually no fungicides because of the effectiveness of the disease-resistance soybean cultivars that have been developed.

The use of classical breeding programs is now being augmented with new plant breeding efforts using transgenic and other genetic engineering techniques. In March 1995, the U.S. Environmental Protection Agency approved for the first time a limited registration of genetically engineered plant pesticides to Ciba and Mycogen Plant Sciences, and in August 1995, granted conditional approval for full commercial use of a transgenic pesticide to combat the European corn borer (Environmental Protection Agency [1995]). This plant pesticides, Bt corn, is produced when the genetic information related to insecticidal properties is transferred from the Bt bacterium to the corn plant. This technology could reduce the need for conventional chemical insecticides in corn production. In 1995, 26 percent of U.S. corn planted acreage was treated with insecticides and corn borer is one of the top insect pests targeted for treatment.

Since these new corn varieties, however, contain natural genes and genes produced from the soil bacteria Bt, many scientists are concerned that the new corn will hasten pest immunity to Bt. That is especially a concern for the growing number of producers who rely on the foliar-applied Bt, and has led the U.S. Environmental Protection Agency to approve the new pesticides conditional on the monitoring for pest resistance and the development of a management plan in case the insects become resistant.

While most classical breeding programs have focused on pests resistant to chemicals or treatments that were too expensive (Zalom and Frye [1992]), consumer concern over pesticides in agricultural products has prompted biotechnology companies to enter the genetically engineered plant market. As agricultural biotechnology products attain commercial success, some private investment funding may shift from the smaller pharmaceutical markets toward agricultural crop protection (Niebling [1995]). On the other hand, consumer acceptance of bioengineered Bt corn, Bt cotton, and other genetically engineered crops has not yet been demonstrated in major U.S. markets. A 1992 survey of consumer attitudes about food biotechnology found that most consumers want information on labels about various food characteristics, including the use of biotechnology (Hoban and Kendall [1993]).

APHIS (Animal Plant Health Inspection Service) has approved or acknowledged 638 field trials for insect-resistant varieties since 1987, 286 field trials to test viral resistance, and 94 field trials for fungal resistance (Economic Research Service [1997]).

BIOPESTICIDE USE IN THE CONTEXT OF PESTICIDE POLICY

Many factors affect the adoption of biopesticides as a crop production technology. Pest cycles and annual fluctuations caused by weather and other environmental conditions often determine whether infestation levels reach treatment thresholds. Changes in farm biopesticide use are related to producers' decisions on the amount and mix of crops to plant. Given this, other factors that influence biopesticide use decisions are relative factor prices and government farm, conservation, and regulatory policies.

(A) RELATIVE FACTOR PRICES

The changing relative prices between different pesticides is important especially as biopesticides strive to replace conventional pesticides. For example, the largest pesticide market in the United States is cotton. It accounts for 35 percent of the insecticide market with about $850 million in sales in 1996 (Economic Research Service [1997]). Bt cotton has the potential to be a challenge to conventional foliar sprays for control of the budworm/bollworm complex. The cost of Bt cotton, including Tracer (DowElanco), Pirate (American Cyanimid), Proclaim (Merck), and Confirm (Rohm and Haas), however, averages about $34 per acre. Conventional insecticide treatment averages around $10.20 per acre. Given the current price differential, Bt cotton is not likely to replace conventional insecticides.

In another example, biological control has been shown to be effective on Canada thistle, leafy spurge, the knapweeds, St. Johnswort, musk thistle, and other weeds. Biological control through beneficial insects (e.g., Canada thistle stem mining weevil and musk thistle rosette weevil), however, is priced substantially higher than, say, atrazine. In 1996, the price of atrazine (a common herbicide used to control these weeds) was approximately $3.90 per acre while biological control is priced at about $70.00 per acre (Economic Research Service [1997]). Thus, biological control will not likely replace atrazine or one of the other triazine herbicides in the near future.

(B) Government Commodity and Conservation Programs

Federal commodity and conservation programs affect agricultural biopesticide use in part through the amount of acres planted. Past commodity programs were designed primarily to provide price and income protection for farmers. Land set-aside requirements helped restrict supply and increase commodity prices (Aspelin [1984]). While the Federal Agriculture Improvement and Reform Act (FAIR) of 1996 eliminated those set-aside requirements, the re-authorized Conservation Reserve Program, designed for environmental objectives, pays producers to keep acreage in conserving uses rather than in production. Fewer acres planted generally implies less biopesticides applied. Other Federal agricultural conservation programs influence production practices on planted acreage, which in turn will affect biopesticide use.

(C) Agricultural Chemical Regulation Implications for Biopesticide Use

Pesticide regulation in its modern form began with the enactment of the Federal Insecticide, Fungicide, and Rodenticide Act (FIFRA) in 1948. Under this mandate, Congress required that all chemicals for sale in interstate commerce be registered against the manufacturers' claims of effectiveness. The law also required manufacturers to indicate pesticide toxicity on the label. Congress amended FIFRA in 1954, 1959, and 1964, but, in practice, pesticide regulation by 1970 meant efficacy testing and labeling for acute (short-term) toxicity. Pesticide regulation passed into a new phase with the 1972 amendment to FIFRA and the transfer of regulatory jurisdiction to the Environmental Protection Agency (EPA). Under this new regulatory regime, Congress gave the EPA the responsibility of re-registering existing pesticides, examining the effects of pesticides on fish and wildlife, and evaluating acute and chronic toxicity. In the 1988 amendment to FIFRA, pesticide producers were required to demonstrate, within 9 years, that all pesticides registered before November 1984 meet current standards (Ollinger and Fernandez-Cornejo [1995]).

Pesticides are also regulated by various provisions of the Federal Food, Drug and Cosmetic Act (FFDCA). Under the FFDCA, the EPA establishes the maximum allowable level (tolerance) of pesticide residues that can be present on foods sold in interstate commerce and the Food and Drug Administration (FDA) monitors food and feed for pesticide residues.

In 1996, Congress passed the Food Quality Protection Act (FQPA) which was intended to update and resolve inconsistencies in the two major pesticide statutes: FIFRA and FFDCA. The major components of the FQPA address the issues of setting a single, health based standard (i.e., a reasonable certainty of no harm) for all pesticides in all foods (although benefits can continue to be considered in certain instances when setting standards), providing special protection for infants and children, regulatory relief for minor use pesticides, expediting approval for safer (reduced-risk) pesticides, requiring periodic re-evaluation of pesticide registrations and tolerances and reauthorizing and increasing registrant fees to fund such reevaluations, establishing national uniformity of tolerances unless States petition for an exception, and mandating the distribution of information in grocery stores on the health risks of pesticides and how to avoid such risks (*Pesticide and Toxic Chemical News* [August 21, 1996], Environmental Protection Agency [1996], Jaenicke [1997]).

The critical components of the FQPA as far as biopesticides are concerned deal with expediting the review of minor use pesticides and expediting the approval of reduced-risk pesticides. Both sections of the legislation should serve to accelerate the development and commercialization of new biological approaches to pest control. EPA is giving high priority to implementing the Minor Use Provisions of FQPA. It has created a new program dedicated solely to coordinating minor use issues within the Office of Pesticide Programs. The definition of a minor use crop has been determined to be: (1) a crop produced on fewer than 300,000 acres or (2) a major crop (a crop grown on more than 300,000 acres) for which the pesticide use pattern is so limited that revenues from expected sales will be less than the cost of registering the pesticide and (a) there are insufficient efficacious alternatives for the use, (b) alternatives pose greater risks, (c) the minor use is significant in managing pest resistance, or (d) the minor use plays a significant part in integrated pest management (IPM). The first part of the definition means that all but 26 of the 600 plus crops produced in the United States are minor crops. EPA will consider every crop in the United States to be a minor crop, except for almonds, apples, barley, canola, carrots, corn (field and sweet), cotton, grapes, hay (alfalfa and other), lettuce, oats, oranges, peanuts, pecans, popcorn, rice, rye, snapbeans, sorghum, soybeans, sugarcane, sugarbeets, tobacco, tomatoes, sunflowers, and wheat.

Provisions intended to help preserve the availability of minor use pesticides include expediting the review of data submitted in support of minor uses, granting time extensions for submitting data on minor uses, and giving those who invest in data development for minor uses additional exclusive rights to use of the data to support registration. The minor use program at EPA in conjunction with a similar program at the U.S. Department of Agriculture will coordinate decisions on minor use issues in consultation with growers. A revolving grant fund is authorized at the U.S. Department of Agriculture to fund the generation of data necessary to support minor use registration. The Department of Health and Human Services is authorized to fund studies in support of registration or reregistration of minor use pesticides that are important for public health purposes.

The Reduced-Risk Pesticide Initiative is designed to encourage the development, registration, and use of new pesticide chemicals which would result in reduced risks to human health and the environment compared to existing alternatives. In 1995, the average amount of time it took to register a new conventional pesticide was 38 months; the new reduced risk pesticides take, on average, on 14 months (Environmental Protection Agency [1997]). Since 1993, 29 new chemical submissions have been received by EPA as reduced risk pesticide candidates. Of the 29, 17 met the reduced risk criteria for expedited review. Nine of those 17 have been registered.

In November 1994, EPA established a separate division in the Office of Pesticide Programs, the Biopesticides and Pollution Prevention Division, to encourage the development of reduced risk pesticides and to manage the registration and reregistration of biopesticides. Biopesticides are defined by EPA to include (1) naturally occurring and genetically engineered microorganisms, (2) genetically engineered plants that produce their own pesticides (such as crops that produce the insecticidal proteins from the Bt bacteria), and (3) naturally occurring compounds, or compounds essentially identical to naturally occurring compounds, that are not toxic to the target pest (such as pheromones). EPA approved 14 new biopesticide active ingredients in fiscal year

1995 and 10 in fiscal year 1996 - representing over one-third of new active ingredients registered in those years. EPA also issued Reregistration Eligibility Decision documents for 8 biopesticides.

FQPA explicitly recognizes the importance of reduced-risk pesticides and supports expedited review to help these pesticides reach the market sooner and replace older and potentially riskier chemicals. The new law defines a reduced-risk pesticide as one which "may reasonably be expected to accomplish one or more of the following: (1) reduces pesticide risks to human health; (2) reduces pesticide risks to non-target organisms; (3) reduces the potential for contamination of valued, environmental resources; or (4) broadens adoption of IPM or makes it more effective."

Other statutes with the potential to affect pesticide use include the Clean Air Act, Clean Water Act, Safe Drinking Water Act, Coastal Zone Management Act (CZMA), and the Endangered Species Act. The Water Quality Act of 1987 (sec. 319) and the CZMA address nonpoint sources of pollution, such as that from farm fields. These are discussed in detail in the *Farm Chemicals Handbook* (Meister Publishing Company [1996]).

Effects of Policies on the Development and Use of Pesticides

In the agricultural sector, policy influences on pesticides take two forms. First, policy affects the choice of production practice which will be relatively more or less pesticide intensive depending on the policy. Second, government policy directly impacts the development of new pesticides and pest management practices. Each of these issues is discussed in turn.

(A) The Direct and Indirect Impact of Government Policy on the Use of Pesticides

Government regulation attempts to mitigate the adverse effects of pesticide use. Due largely to concerns over the environmental problems associated with pesticide use in production agriculture, farmers, with some involvement on the part of government in the form of costs sharing and educational and technical assistance, are experimenting with a myriad of new and traditional tools, materials and practices. Short-term productivity enhancement and chemical use efficiency are the major goals in some of the systems, while environmental risk reduction and the long run sustainability of farming are more prominent in others. To help minimize the trade-off between lower pesticide use and reduced net farm income, IPM, precision farming, and cultural and biological pest and nutrient management methods are among options encouraged by some. The first two attempt to increase efficiency of chemical use, while the latter two attempt to avoid chemical use altogether.

Pest scouting, economic thresholds, and other tools to help the farmer determine when to make pesticide applications, which pesticides to use, and how much to use have been developing for decades, and "expert systems" and other decision support systems to integrate these elements are emerging. Precision farming and herbicide-resistant bioengineered crops are efficiency technologies that aim at reducing pesticide use that are just now being developed and commercialized.

Scouting and threshold use is widespread in specialty crop production (Vandeman et al. [1994]). Half of the U.S. fruit and nut acreage and nearly three-quarters of the vegetable acres in the surveyed States were scouted for insects, mostly by professional scouts. Growers reported using thresholds as the basis for making pesticide treatment decisions on virtually all of these scouted acres. Potato growers reported that 85 percent of their acreage was scouted and thresholds were used in making nearly three-quarters of their insecticide application decisions.

A number of alternative production practices - including crop rotation, conservation tillage, alterations in planting and harvesting dates, trap crops, sanitation procedures, irrigation techniques, fertilization, physical barriers, border sprays, cold air treatments, and habitat provision for natural enemies of crop pests - are now being relatively more extensively used for managing crop pests. Their diffusion is expected to grow more widespread. These alternative production practices work by preventing pest colonization of the crop, reducing pest populations, reducing crop injury, and enhancing the number of natural enemies in the cropping system (Ferro [1996]).

Crop rotation is one of the most important of these techniques that is currently in widespread use. Over half of the corn and soybeans were grown in rotation with each other during mid-1990s (Lin and Delvo [1994]). Farmers rotating corn with other crops used insecticides less frequently than did those planting corn two years in succession (11 percent versus 46 percent). Corn is often grown as a monocrop in areas which have high demand for livestock feed, and where climatic restrictions limit the soybean harvest period (Edwards and Ford [1992]). Crop rotation is much less prevalent for cotton, however, which has among the highest per acre returns of the field crops grown in the U.S., and less than one-third of the cotton producers use this technique.

(B) THE IMPACT OF GOVERNMENT POLICY ON THE DEVELOPMENT OF NEW PESTICIDES

Research and development expenditures are an important factor affecting the development of new pesticides. There is a fairly extensive literature exploring the relationship between research and development expenditures on a good or service and the use of that good or service (Kahn [1971], Scherer [1980], and Office of Technology Assessment [1986]). Basically, research and development expenditures in basic industries, such as the chemical industry, are conditioned by the present value of expected net revenue. That is, research and development expenditures are made with the expectation that a profit for the firm will result. The net present value is a function of costs such as research and development expenditures and regulatory costs of getting a new pesticide registered as well as the revenue generated from the sale of the pesticide.

In this setting, and as apparent from the foregoing discussion, pesticide policy is only one, albeit an important one, of the factors affecting pesticide use and hence the development of biopesticides. Any government policy affecting relative factor prices, price responsiveness, expected returns from chemical use, conservation, and technology development and adoption will impact the development of new pesticides.

The two major statutes, FIFRA and the FFDCA instruct regulators to weigh the benefits of pesticide use against "unreasonable" risks. This balancing process has been characterized such that the use of the term 'unreasonable risk' implies that some risks will be tolerated under FIFRA, it is clearly expected that the anticipated benefits will outweigh the potential risks when a pesticide is used according to commonly recognized, good agricultural practice (National Research Council [1993]).

A study of the impact of pesticide regulation on innovation and the market structure in the U.S. pesticide industry shows that pesticide regulation in the United States has encouraged the introduction of fewer, yet less toxic pesticides (Ollinger and Fernandez-Cornejo [1995]).[19] The 1972, 1978, and 1988 amendments to FIFRA require that new and existing pesticides meet strict health and environmental standards. Requirements for pesticide registration with the EPA include field tests that can include up to 70 different types of tests that can take several years to complete and cost millions of dollars. They consist of toxicological studies, a two-generation reproduction and teratogenicity study, a mutagenicity study, oncogenicity studies, and chronic feeding studies. The toxicological studies include acute (immediate), subchronic (up to 90 days), and chronic (long-term) effects. Other tests are used to evaluate the effects of pesticides on aquatic systems and wildlife, farmworker health, and environmental fate. Recent estimates suggest that research and development of a new chemical pesticide (including the testing indicated above) costs between $50-70 million and takes 11 years (Ollinger and Fernandez-Cornejo [1995]). As a consequence of the regulation requirements, pesticide firms refocused their research away from persistent and toxic pesticides. The number of pesticides with chronic (long-term) toxicity dropped by 86 between the 1972-76 and 1987-91 periods and lower toxicity pesticides account now for more than half of the pesticide sales. A 10 percent increase in testing costs is associated with a 2.8 percent increase in the proportion of "less toxic" pesticides registered. In 1996, the Office of Pesticide Programs of the Environmental Protection Agency registered 22 new pesticide active ingredients, more than half of which were considered reduced-risk pesticides. These decisions included the approval of ten biopesticides and twelve new chemicals, which include three reduced-risk chemicals (Office of Pesticide Programs [1996]). The biopesticides include Bt Cotton (Monsanto), 1-octen-3-ol (Armatron), Jojoba oil (IJO Products), Bt (CRYMAX) (Ecogen, Inc.), Myrethecium verrucaria (Abbott Laboratories), meat meal (Lakeshore Enterprises), red pepper (Lakeshore Enterprises), Verticillium lecanii (Abbott Laboratories), NK Bt corn (Northrup King), Monsanto Bt corn (Monsanto), and Lavandin oil (S.C. Johnson and Sons).

Pesticide regulation has also had undesirable consequences. Regulation discouraged new chemical registrations: the number of new pesticides registered by the EPA in 1987-91 was half that of 1972-76 and each 10 percent increase in pesticide regulatory costs caused a 2.7-percent reduction in the

[19] High toxicity pesticides are those that belong to Class I acute toxicity (indicated on the label), or are chronically toxic to humans, or fish and wildlife. Lower toxicity pesticides are all others (Ollinger and Fernandez-Cornejo [1995]).

number of new pesticides introduced (Ollinger and Fernandez-Cornejo [1995]). The higher regulatory costs contributed to an industry-wide increase in research spending which encouraged some small firms to leave the pesticide industry. Pesticide regulation also encouraged firms to focus their research on pesticides used in larger crop markets such as corn and soybeans abandoning minor crop markets, such as horticultural crops. The decline in new registrations of chemical pesticides suggest that there are market opportunities for biopesticides and genetically modified plants. These products are not only environmentally preferable but also less costly to develop and register than chemical pesticides. Thus, it has been estimated that the average cost of developing a biopesticide ranges from three million to five million dollars versus $50 to $80 million for the development of conventional pesticides (CRC Press [1996]). Such new products, however, as noted previously are only effective against a narrow range of pests. The 15 largest (in terms of retail sales) agricultural chemical companies are developing biopesticides including pheromones, bioinsecticides (viruses), botanical extracts, soybean seed, corn and sorghum seed, microbial products, Bt manufacturing, microsponge formulation, and gene insertion.

CONCLUSION

Biopesticides developed and used in the future will emerge against the backdrop of the environmental effects associated with the use of conventional pesticides and government policies designed to control these effects. In the final analysis, farmers' choices on pesticides will be influenced by the prevailing costs and benefits of conventional pesticides and their alternatives including biopesticides.

The outlook for pesticide use is complicated, though some directions can be perceived. There are a number of factors that will serve potentially to impact pesticide use which in turn will affect the development of biopesticides. These include pesticide regulation, the FAIR Act, the crops planted, the management of ecologically-based systems, and consumer demand for "green" products.

(A) Pesticide Regulation

Pesticide regulation will continue to exert a major influence on pesticide use and the development of pesticides and pest management alternatives in the United States. The number of pesticide active ingredients for sale in the U.S. has decreased by 50 percent since 1989 due to EPA's reregistration process (Pease et al. [1996]). Moreover, regulatory changes involving the removal from the market of pesticides, which had been previously registered, because of evidence on unacceptably high health hazards from occupational exposure may also undermine the confidence that farmers have in the safety of pesticide use (Bender [1994]). Finally, implementation of the Food Quality Protection Act of 1996 will impact pesticide registrations in general and biopesticides in particular. The FQPA is designed to expedite the review of minor use pesticides and expediting the approval of safer pesticides. Both sections of the legislation should serve to accelerate the development and commercialization of new biological approaches to pest control. Although the legislation does not expressly recognize the presumption of biologicals for the expedited review category, the provision will assist in promoting new research and development activities (Environmental Protection Agency [1997]).

(B) The FAIR Act

Short-term effects of the Federal Agriculture Improvement and Reform (FAIR) Act on biopesticide use stem from greater flexibility provided to producers through elimination of base acreage and set-asides. Elimination of base acreage will facilitate rotations, which could reduce insecticide use. Without the concern of maintaining base acreage to receive Federal deficiency payments, one would expect producers to plant crops for which returns are higher (e.g., corn rather than wheat) where producers have such options. Hence, biopesticide use will change based on how and where the mix of crops will change. For example, if producers plant more corn, a more chemically-intensive crop, rather than wheat, which generally requires less chemicals, one would expect chemical use to increase (Economic Research Service [1997]). Elimination of set-asides, other acreage reduction programs, and a potentially smaller Conservation Reserve Program could result in increased planted acreage with resulting increased biopesticide use. Set-asides, however, have been relatively low if not zero for several program crops recently so the increase in planted acreage would not be dramatic. In net, greater biopesticide use would be expected if more chemically-intensive crops are planted on existing acreage and greater acreage overall is planted, but greater crop rotations would curb such growth (Office of Pesticide Programs [1996]).

In the long run, input use will hinge on the relative marginal productivity and cost among labor, pesticides, fertilizer, and other factors, which the FAIR Act will not alter. As nominal prices decline from 1995/96 peaks, and real prices are anticipated to continue to decline, there will be less incentive to apply inputs whose value of the marginal product does not increase. While Federal support will fall, market demand is expected to keep commodity prices relatively high by historical standards. Thus market incentives will replace Federal incentives regarding application of chemical inputs.

(c) Crops Planted

The U.S. Department of Agriculture projects that crop acreage of eight major crops will rise between 5 and 10 percent by the year 2005 from 1995 levels (World Agriculture Outlook Board [1997]).[20] Both corn and wheat acreage are expected to rise by about 10 million acres each, while that of cotton is expected to decline by about 3 million. Hence, while the changing mix of crop acreage complicates a projection of pesticide use, it seems likely that from increased planted acreage, biopesticide use will continue to rise, other influences held constant (Economic Research Service [1997]).

(d) Ecologically Based Management Systems

The U.S. Department of Agriculture announced several years ago that switching to an ecosystem-based approach for managing natural resources is among its major agricultural priorities for the 1990s (Comanor and Gelburd [1994]). The new approach, which is to be adopted gradually, adds the goal of maintaining or improving the condition of the land as the context for providing sustainable levels of goods and services. Ecosystem management is partly based on the emerging research on biodiversity, ecosystems, and environmental accounting from new scientific disciplines such as conservation biology, landscape ecology, and ecological economics. The impacts of species loss on crop breeding programs and more complete accounting of the costs of food and fiber production are some of the issues that are addressed.

While much of the initial application of this approach has been for national forests and other public resources, its potential use for crop production

---

[20] Crops are barley, corn, upland cotton, oats, rice, sorghum, soybeans, and wheat.

systems is also being explored. The National Research Council [1995] has published the results of a study to examine ecologically based pest management practices for agriculture and forestry. The NRC report concludes that pest resistance and other problems created by pesticide use has created a need for an alternative approach to pest management that can complement and partially replace current chemically based pest-management practices. Ecosystem based pest management is the approach that was recommended.

U.S. Department of Agriculture's Forest Service adopted an ecosystem management philosophy in June 1992, and this approach has been influencing the development of forest management plans in the Pacific Northwest and other areas. For example, ecosystem design - arranging landscape structures in the watershed to support species biodiversity as well as timber production and recreation - was used in the recently developed forest management plan for a 30,000-acre watershed in Mt. Hood National Forest (Pease et al. [1996]).

A number of states have begun to examine ecosystem-based pest management in specific agricultural cropping systems. Maine's Agricultural and Forest Experiment Station, for example, has recently reported results from the first four years of its pioneering industry-supported ecosystem project on sustainable potato production (University of Maine [1996]). Various states and regions also have some ecosystem research underway, including some federally-funded IPM projects which are looking at biological alternatives, as well as public-private efforts at the local level such as the Chesapeake Bay watershed project and BIOS project for California almond growers.

(E) Consumer Demand for "Green" Products and the Market Response

The market for food products with a green label - such as certified organic and IPM - has been growing in the United States. While organic food products only account for about 1 percent of total retail food sales at the present time, overall organic sales reached $2.8 billion in 1995, and have increased over 20 percent annually since 1989 (Natural Food Merchandiser [1996]). Organic foods are becoming more widely available to U.S. consumers through the growing number of large natural food stores, mainstream supermarkets, and direct outlets such as farmers markets.

A consistent set of national standards for organic production and processing, which is currently being developed by USDA, is expected to enhance consumer confidence in the United States. Development of these standards was required by The Organic Foods Production Act, which was passed by Congress as part of the Food, Agriculture, Conservation, and Trade Act of 1990. This legislation requires that all except the smallest organic growers will have to be certified by a state or private agency accredited under the national standards.

The National Organic Standards Board, which was appointed by the U.S. Department of Agriculture to help implement the provisions of the Act, currently defines organic agriculture as a sustainable production management system that promotes and enhances biodiversity, biological cycles, and soil biological activity. It is based on minimum use of off-farm production inputs and on practices that maintain organic integrity through processing and distribution to the consumer (Ricker [1997]).

In tandem with the growth in demand for food with a green label is the demand for food with less pesticide residue. Biopesticides are viewed as being safer than chemical products (CRC Press [1996]). Early in their development, biopesticide companies - including biosys, Consep, Ecogen, and Mycogen - were forced to concentrate their marketing efforts on niche markets (primarily vegetables and fruits) because of the mediocre performance of their products. They are now establishing themselves in major markets like cotton and corn. Biopesticide companies concentrated on niche markets because products did not have the price/performance (efficacy) characteristics necessary to compete in the larger pesticide markets. They found the less competitive niche markets were a sheltered corner where they could mature. These small markets gave biopesticides a sales base for products that did not have the attributes desires in larger markets. Biopesticide companies have since invested heavily in upgrading products so they can move beyond their niche market base into major pesticide markets (Bioscience Securities, Inc. [1996]).

Biopesticide companies have made substantial improvements in recent years and have become more competitive with conventional chemical companies. Most biopesticides have improved price/performance characteristics due to improved technology which has resulted in better efficacy and lower costs. For example, recently developed Bt products including, Bt (CRYMAX) (Ecogen) and Mattch (Mycogen) have more concentrated toxins, give more consistent performance, have longer residual formulations, and contain relatively more potent Bt strains. These improvements have allowed the companies to lower prices.

Finally, companies have repositioned products in markets where the products add value and have focused on educating growers on the use of the products. For example, nematodes have been repositioned to address selected citrus and ornamental markets. These sorts of repositioning - focusing on selected markets - should continue in the future as companies recognize where their products are most effective.

REFERENCES

Abler, D.G., "Issues in Pesticide Policy," *Northeastern Journal of Agricultural and Resource Economics*, Vol. 21 (1992), pp. 93-94.

Aspelin, A.L., Pesticide Industry Sales and Usage: 1992 and 1993 Market Estimates, *BEAD*, Office of Pesticide Programs, U.S. Environmental Protection Agency, Washington, DC, June 1984.

Bender, J. *Future Harvest: Pesticide-free Farming*, University of Nebraska Press, Lincoln, 1994.

Bioscience Securities, Inc., *The Outlook for Biopesticides*, Bioscience Securities, Inc., Orinda, CA, 1996.

Carlson, G.A., and E.N. Castle, "Economics of Pest Control," *Pest Control Strategies for the Future*, National Academy of Sciences, Washington, 1972.

Comanor, S., and J. Gelburd, "An Ecosystem Approach to Resource Management," *Agricultural Outlook*, AO-204, Economic Research Service, U.S. Department of Agriculture, January-February, 1994.

CRC Press, "Future Development of Biopesticides Expedited by FQPA," *Pesticide and Toxic Chemical News*, Vol. 24 (October 9, 1996), pp. 19-20.

Economic Research Service, *Agricultural Resources and Environmental Indicators*, U.S. Department of Agriculture, Washington, DC, 1997.

Edwards, C.R., and R.E. Ford, "Integrated Pest Management in the Corn/Soybean Agroecosystem," in *Food, Crop Pests, and the Environment*, F. G. Zalom and W. E. Fry, eds., The American Phytopathological Society, St. Paul, MN, 1992.

Environmental Protection Agency, "EPA Issues Conditional Approval for Full Commercial Use of Field Corn Plant Pesticide Targeting the European Corn Borer," *EPA Press Release*, Washington, DC, 1995.

Environmental Protection Agency, Major Issues in the Food Quality Protection Act of 1996, Environmental Protection Agency, *Prevention, Pesticides, and Toxic Substances*, Washington, DC, August 1996.

Environmental Protection Agency, 1996 Food Quality Protection Act - Implementation Plan, Environmental Protection Agency, *Prevention, Pesticides, and Toxic Substances*, Washington, DC, 1997.

Ferro, D.N., "Cultural Controls," *Electronic IPM Textbook*, E.B. Radcliffe and W. D. Hutchison, eds., University of Minnesota and the Consortium for International Crop Protection, Minneapolis, MN, 1996.

Hoban, T., and P. Kendall, *Consumer Attitudes about Food Technology: Project Summary*, Extension Service, North Carolina State University, Raleigh, NC, 1993.

Hunter, C., Suppliers of Beneficial Organisms in North America, *PM 94-03*, California Environmental Protection Agency, Department of Pesticide Regulation, Sacramento, CA, 1994.

Jaenicke, E., *The Myths and Realities of Pesticide Reduction*, Henry A Wallace Institute for Alternative Agriculture, Beltsville, MD, 1997.

Kahn, A., *The Economics of Regulation*, John Wiley and Sons, Inc., New York, 1971.

Landis, D.A., and D.B. Orr, "Biological Control: Approaches and Applications," *Electronic IPM Textbook*, E.B. Radcliffe and W. D. Hutchison (eds.), University of Minnesota and the Consortium for International Crop Protection, Minneapolis, MN, 1996.

Lin, B.H., and H. Delvo, *Pest Management Practices on 1993 Corn, Fall Potatoes, and Soybeans*, Natural Resources and Environment Division, Economic Research Service, U.S. Department of Agriculture, Washington, 1994.

Meister Publishing Company *Farm Chemicals Handbook*, Willoughby, OH, 1996.

National Research Council, *Pesticides in the Diets of Infants and Children*, National Academy Press, Washington, DC, 1993.

National Research Council, *Ecologically Based Pest Management: New Solutions for a New Century*, National Academy Press, Washington, DC, 1995.

Natural Foods Merchandiser, "Widening Market Carries Organic Sales to $2.8 Billion in 1995," *Natural Foods Merchandiser*, Vol. 17 (1996), pp. 5-7.

Niebling, K., "Agricultural Biotechnology Companies Set Their Sights on Multi-Billion $$ Markets," *Genetic Engineering News*, July 1995.

Office of Pesticide Programs, *Office of Pesticide Programs Annual Report for 1996*, Environmental Protection Agency, Office of Pesticide Programs, Washington, DC, November 1996.

Ollinger, M., and J. Fernandez-Cornejo, Regulation, Innovation, and Market Structure in the U.S. Pesticide Industry, *Agricultural Economic Report 719*, U.S. Department of Agriculture, Economic Research Service, June 1995.

Pease, W.S., J. Liebman, D. Landy, and D. Albright, *Pesticide Use in California: Strategies for Reducing Environmental Health Impacts*, California Policy Seminar, University of California, Berkeley, Center for Occupational and Environmental Health, 1996.

Poritz, N., "Biological Control of Weeds," *Biological Control of Weeds 1996*, Montana State University Press, Bozeman, MT, 1996.

Ricker, H.S., "The National Organic Program - Status and Issues," *Proceedings of the Third National IPM Symposium/Workshop*, U.S. Department of Agriculture, Economic Research Service, Washington, DC, 1997.

Ridgeway, R., M. Inscoe, and K. Thorpe, *Biologically Based Pest Controls: Markets, Industries, and Products*, U.S. Department of Agriculture, Agricultural Research Service, Washington, DC, 1994.

Scherer, F.M., *Industrial Market Structure and Economic Performance*, Rand McNally College Publishing Company, Chicago, 1970.

U.S. Congress, *Office of Technology Assessment, Technology, Public Policy, and the Changing Structure of American Agriculture, OTA-F-285*, U.S. Government Printing Office, Washington, DC, March 1986.

U.S. Congress, *Office of Technology Assessment, Biologically Based Technologies for Pest Control, OTA-ENV-636*, U.S. Government Printing Office, Washington, DC, 1995.

U.S. Department of Agriculture, *Agricultural Chemical Usage: Fruit Crops Summary*, U.S. Department of Agriculture, National Agricultural Statistics Service and the Economic Research Service, Washington, various.

U.S. Department of Agriculture, *Agricultural Chemical Usage: Vegetable Crops Summary*, U.S. Department of Agriculture, National Agricultural Statistics Service and the Economic Research Service, Washington, various.

University of Maine, *The Ecology, Economics, and Management of Potato Cropping Systems: A Report of the First Four Years of the Maine Potato Ecosystem Project*, Bulletin 843, Maine Agricultural and Forest Experiment Station, Orono, ME, April 1996.

Vandeman, A., J. Fernandez-Cornejo, S. Jans, and B.H. Lin, *Adoption of Integrated Pest Management in U.S. Agriculture, AIB-707*, U.S. Department of Agriculture, Resources and Technology Division, Economic Research Service, 1994.

World Agricultural Outlook Board, *Agricultural Baseline Projections to 2005, Reflecting the 1996 Farm Act, WAOB-97-1*, U.S. Department of Agriculture, Washington, DC, February 1997.

Zalom, F., and W. Fry, Food, *Crop Pests, and the Environment*, The American Phytopathological Society, St. Paul, MN, 1992.

CHAPTER 6

THE CHANGING PATTERN OF SOIL EROSION IN THE UNITED STATES

INTRODUCTION

Soil erosion remains a serious environmental problem in parts of the United States, even after more than 60 years of state and federal efforts to control it. The most widespread offsite erosion-related problem is impairment of water resource use (National Research Council [1993] and Natural Resources Conservation Service [1996]). The Environmental Protection Agency has identified siltation associated with erosion in rivers and lakes as the second leading cause of water quality impairment, and agricultural production is identified as the leading cause of water quality impairment (Environmental Protection Agency [1995]).

Related causes of water use impairment are sedimentation and eutrophication. When soil particles wash off a field, they may be carried in runoff until discharged into a water body or stream. Not all agricultural constituents that are transported from a field reach water systems, but a significant portion does, especially the more chemically active, finer soil particles. Once agricultural pollutants enter a water system, they lower water quality and can impose economic losses on water users. These offsite impacts can be substantial. The offsite impacts of erosion are potentially greater than the onsite productivity effects in the aggregate (Foster and Dabney [1995]). Therefore, society may have a larger incentive for reducing erosion than farmers have.

If the runoff reaches the water body or stream, soil particles can be suspended in the water, or settle out as sediment, depending on the velocity of the waterflow and the size of the soil particles. In each case, water use can be affected.

Suspended soil particles affect the biologic nature of water systems by reducing the transmission of sunlight, raising surface water temperatures, and affecting the respiration and digestion of aquatic life. The effects on aquatic life, and the reduction in aesthetic quality of recreation sites, can reduce the value of water for recreation uses. Suspended soil particles impose costs on water treatment facilities which must filter out the particles. Suspended soil particles can also damage moving parts in pumps and turbines.

Even when soil particles settle on the bottom of a river or lake, they can cause serious problems for aquatic life by covering food sources, hiding places, and nesting sites. Sedimentation can clog navigation and water conveyance systems like roadside ditches, reduce reservoir capacity, and damage recreation sites. In streambeds, sedimentation can lead to an increase in the frequency and severity of flooding by reducing channel capacity.

Wind erosion produces offsite impacts that can be as dramatic as the Dust Bowl of the 1930s. It has not, however, received the attention given to the more widespread water erosion impacts. Damage can include higher maintenance of buildings and landscaping, pitting of automobile finishes and glass, greater wear on machinery parts, increased soiling and deterioration of retail inventories, costs of removing blown sand and dust from roads and ditches, and increased respiratory and eye disorders. Offsite damages from wind erosion depend on the extent and location of population centers relative to prevailing winds and wind erosion sources (Piper and Lee [1989]). Consequently, damage estimates for one area cannot readily be extrapolated to other areas, nor can the impact of wind erosion from cropland or other agricultural land be differentiated from wind erosion originating on nonagricultural land.

Offsite impacts of both sheet and rill (water) erosion and wind erosion may be subject to threshold effects (Zison et al. [1977]).21 A reduction in erosion may not produce proportional improvements in water or air quality unless they are quite large in relation to total loads. In economic terms, the costs of

---

[21] Sheet and rill erosion is the most common form of agricultural soil erosion. It occurs when raindrops or irrigation detach soil particles from the soil surface and transport them in thin sheets of water moving across unprotected slopes. As runoff water becomes concentrated first into rills and then into separate channels, it begins to cut gullies, removing larger volumes of soil.

erosion control practices that result in only small reductions in erosion may produce few, if any, offsite benefits.

A third and somewhat ancillary erosion-related problem deals with wildlife. Monoculture production and field consolidation have diminished habit diversity in areas where agriculture once contributed to diversity (Strohbehn [1986]). Soil conservation practices frequently enhance wildlife habitat. Field borders, windbreaks, hedgerows, riparian buffers, and wildlife habitat management can increase habitat diversity. Practices aimed at wildlife protection, however, often divert land from row crop production thereby creating opportunity costs.

SOIL CONSERVATION PROGRAMS

Soil conservation policies have existed in the United States for more than 60 years. Initially, these policies focused on the on-farm benefits of keeping soil on the land and increasing net farm income. The conservation and related water quality programs administered by the U.S. Department of Agriculture (USDA) primarily have been designed to induce the voluntary adoption of conservation practices. The USDA has used a number of policy tools including on-farm technical assistance and extension education, cost-sharing assistance for installing practices preferred, rental and easement payments to take land out of production and place it in conservation uses, R&D for developing, evaluating and improving conservation practices and programs, and compliance provisions that require the implementation of specified conservation practices or the avoidance of certain land use changes if a farmer wants to be eligible for Federal agricultural program payments. Regulatory and tax policies have not been part of the traditional voluntary approach of U.S. conservation programs. The USDA policy has been to decrease government involvement in farm operations (Reichelderfer [1990]).

The USDA conservation programs are closely tied to State and local programs. Federal and State agencies cooperate with a system of special-purpose local (county) conservation districts that are authorized by State law to provide education and technical assistance to farmers, and county Agricultural Stabilization and Conservation (ASC) committees to handle cost-sharing

(Libby [1982]).[22] The system assures that financial support and technical assistance are focused on the set of problems relevant to the geographic region and the national interest. The adoption of an alternative production practice generally does not occur as a consequence of any one specific assistance program (Missouri MSEA [1995]). The USDA has a memorandum of understanding with each conservation district to assist in carrying out a long run term program.[23] Conservation districts have proven to be practical organizations through which local farmers and the Federal Government can join forces to carry out needed soil conservation practices (Rasmussen [1982]). The demand for information has changed over time. Not long ago, an extension agent was the primary source of information on new technologies (van Es [1984] and Hefferman [1984]). Now, however, farmers are relying on many additional sources of information, including newspapers, magazines, agrichemical dealers, crop consultants, and the Internet.

The U.S. Department of Agriculture Natural Resources Conservation Service (NRCS), formerly the Soil Conservation Service, provides technical assistance to farmers and other land users, including local, State, and Federal agencies that manage publicly owned land. NRCS helps district supervisors and others to draw up and implement conservation plans.

Providing Federal cost-sharing assistance to farmers for voluntary installation of approved conservation practices is the responsibility of State and county ASC committees. Through the Agricultural Conservation Program (ACP), funds were allocated among the States through the State ASC committees on the basis of soil and water conservation needs. ACP practices eligible for cost sharing were established by a national review group representing all USDA agencies with conservation program responsibilities, the Environmental Protection Agency, and the Office of Management and Budget. The practices were designed to help prevent soil erosion and water pollution from animal wastes or other nonpoint sources, to protect the productive capacity of

---

[22] The supervision of these committees was transferred to the Farm Service Agency with the passage of the Federal Agricultural Improvement and Reform Act of 1996. The name of the committees was changed to State Technical Committees.

[23] Many conservation programs to be implemented at the State and local levels require States to submit plans or project proposals and funding needs for Federal approval before actual funds are transferred. For multiyear projects, annual plans of work and documentation of progress are required to receive continued funding. A summary of State programs for erosion control is provided in Magleby et al. [1995].

farmland, to conserve water, to preserve and develop wildlife habitat, and to conserve energy (Holmes [1987]).[24]

The Secretary of Agriculture can also target critical resource problem areas for financial and technical assistance based on the severity of the problem and the likelihood of achieving improvement. Highly erodible and/or environmentally sensitive cropland has recently been targeted because the greatest net social benefits were expected to be associated with a reduction in soil erosion on these land classifications. Targeting, however, will not guarantee that the net benefits (public and private) of any conservation practice will be positive, because net benefits are a function of site-specific factors.

The policy to take land out of production and place it into conservation uses was first used in the Soil Bank Program of the 1950s, and has been significantly increased in the current Conservation Reserve Program (CRP). The Conservation Reserve Program provides for the USDA to enter into 10-15 year agreements with owners and operators to remove highly erodible and other environmentally sensitive cropland from production. Along with conservation, the CRP originally had a second objective of reducing surplus crop production (Osborn [1996]). The more recent emphasis on CRP, however, has been to provide environmental benefits rather than to control the supply of commodities.

Most of the highly erodible land (HEL) (see the Appendix to Chapter 8) contracted into the CRP had suffered much erosion, organic matter loss, and structural deterioration while it was in cultivated crop production. When lands are returned to grass, their structure and organic matter improve and tend to approach the structure and organic matter content of the original grassland soils (Gebhart et al. [1994]). The degree of soil improvement from 10 years of grass is a function of site-specific factors. As a general rule, the greater the amount of soil structure deterioration from past cultural practices, the more likely that grass management will improve the soil's characteristics. Rasiah and Kay [1994] found that if soils had higher levels of organic matter and other stabilizing materials at the time of grass introduction, the time required for soil structure regeneration was reduced. Soils in the CRP typically fit into the category of degraded soils whose organic matter is lower than that of surrounding soils, because they were primarily allowed into the program based

[24] Authority for ACP was terminated on April 4, 1996, when its functions were subsumed by the Environmental Quality Incentives Program (EQIP).

on their highly erodible classification (Lindstrom et al. [1992] and Barker et al. [1996]).

CRP contracts are beginning to expire, so farmers have the option to return the land to crop production. For land that would be returned to production, the improvements in soil quality and erosion reduction gained during the CRP contracts will be rapidly lost if conventional tillage is used. A concern is whether it will be possible to maintain the benefits derived from 10 years of grass. There are several options for post-CRP land.

1. Keep HEL land in the CRP. This would allow the soil in the program to continue to improve and maintain the erosion benefits.
2. Subsidize a rotation that involves, say two to four years in grass production followed by two to four years in grain production.
3. Lower CRP payments to keep the land in grass but allow grazing or haying on the land. Proposals, however, permitting haying and grazing on CRP lands have always been met with considerable opposition by farmers and ranchers who have land already in hay production. They object to subsidized hay production that would compete with their commodity (Schumacher et al. [1995]).

The focus of current conservation research has been on the development of environmentally acceptable and sustainable production practices. Goals of this research are to gain a better understanding of how different soils respond to tillage, what amount of tillage is necessary for optimum crop growth, and what combination of mechanical, chemical, and biological practices are needed to create environmentally sustainable crop production. Many conservation practices have been evaluated by the land grant university experiment stations (Moldenhauer et al. [1994]). USDA has funded several major surveys to provide data to assess the extent and determinants of adoption for particular production practices across a wide range of crops and regions. A more in-depth discussion of conservation-related R&D can be found in Karlen [1990].

A major shift in U.S. conservation policy came in the Food Security Act of 1985 in the form of conservation compliance (Heimlich [1991]). This provision provided farmers with a basic economic incentive to adopt conservation tillage practices or another acceptable plan because agricultural program payments were linked to the adoption of an acceptable conservation system on highly erodible land. While meeting the conservation provisions remains voluntary, a farmer who wants to receive certain agricultural program payments and whose cropland is designated as HEL has no choice but to implement an acceptable conservation plan (Crosswhite and Sandretto [1991]). The conservation compliance provision was innovative because it linked farm program payments (private benefits) to conservation performance (social benefits).[25] In 1982, cultivated HEL accounted for almost 60 percent of the total erosion on U.S. cropland in terms of tons per acre per year per year while it accounted for only 40 percent of total planted acreage (Magleby et al. [1995]). Requirements for conservation compliance were applied to HEL previously cultivated in any year between 1981 and 1985. Conservation compliance required farmers producing crops on HEL to implement and maintain an approved soil conservation system by 1995.

The most recent manifestation of agricultural program policy is the Federal Agriculture Improvement and Reform (FAIR) Act of 1996. It modifies the conservation compliance provisions of the Food Security Act of 1985 to provide farmers with greater flexibility in developing and implementing conservation plans, in self-certifying compliance, and in obtaining variances for problems affecting application of conservation plans. Producers who violate conservation plans, or fail to use a conservation system, on highly erodible land risk loss of eligibility for many payments including production flexibility contract payments. An important aspect of this Act is that in self-certifying compliance, there is no requirement that a status review be conducted for producers who self-certify (Nelson and Schertz [1996]). The FAIR Act also does not differentiate between previously cultivated and uncultivated land thereby eliminating the sodbuster program.[26] Newly cropped highly erodible land may use conservation systems other than the systems previously required under the sodbuster program. Additionally, the FAIR Act established a new program, the Environmental Quality Incentives Program (EQIP), that incorporated the functions of ACP and some other environmental programs, designed to en-

---

[25] Technically, the Food Security Act of 1985 is not the first instance of recognizing the off-site damages of soil erosion and hence the need to target conservation programs. The Soil Conservation Service in 1981 moved to target an increasing proportion of soil erosion programs to areas of high erosion rates in order to reduce substantial off-site damages, and the Agricultural Stabilization and Conservation Service began targeting its Agricultural Conservation Program (ACP) in 1982. The success of these efforts is assessed in Nielson [1985]. Targeting became a general policy instrument, however, with the passage of the Food Security Act of 1985.

[26] Note that the sodbuster program was applicable to HEL uncultivated between 1981 and 1985.

courage farmers to adopt production practices that reduce environmental and resource problems. The acceptable plans will improve soil, water, and related natural resources including grazing lands, wetlands, and wildlife habitat. EQIP must be carried out to maximize environmental benefits provided per dollar expended. During 1996-2002, the Secretary of Agriculture will provide technical assistance, education, and cost-sharing to producers who enter into 5- to 10-year contracts specifying EQIP conservation programs

An important question is just how effective have these programs been in reducing soil erosion. This issue is examined in the next section.

TRENDS IN SOIL EROSION

(A) DATA

The Rural Development Act of 1972 (Public Law 92-419) directed the Secretary of Agriculture to carry out a land inventory and monitoring program that reflects soil, water, and related resource conditions at not less than five year intervals. The Soil and Water Resources Conservation Act of 1977 (Public Law 95-192) stipulated that this inventory be the basis for developing a soil and water conservation program - the National Conservation Program. These inventory activities are conducted within the Natural Resources Conservation Service (formerly the Soil Conservation Service) of the U.S. Department of Agriculture and culminate in the National Resources Inventory (NRI).

For over 60 years, the Natural Resources Conservation Service (Soil Conservation Service) has been involved in conducting periodic inventories of the soil, water, and related resources in the United States. The earliest efforts were reconnaissance studies, the Soil Erosion Inventory of 1934, and the 1945 Soil and Water Conservation Needs Inventory. The 1958 and 1967 Soil and Water Conservation Needs Inventories were the first efforts to collect data nationally for scientifically selected field sites. The Potential Cropland Study of 1975 and the National Resources Inventory of 1977 were extensions and modifications of these earlier inventories. The 1977 NRI was the first inventory to quantify scientifically the extent of erosion occurring on all nonfederal land throughout the United States. The 1982 NRI was the most comprehensive study ever conducted (up to that time) and was designed to obtain natural resource data usable for analysis at a substate (multi-county) level. It contained information on over 365,000 primary sampling units (PSUs).[27] Samples were selected using standard statistical techniques of stratification, area sampling, and clustering. This sampling technique was used on all subsequent inventories. It will serve as the benchmark in analyzing trends in soil erosion in the United States.[28]

Figure 6.1. Total Erosion on Total Cropland (billions of tons)

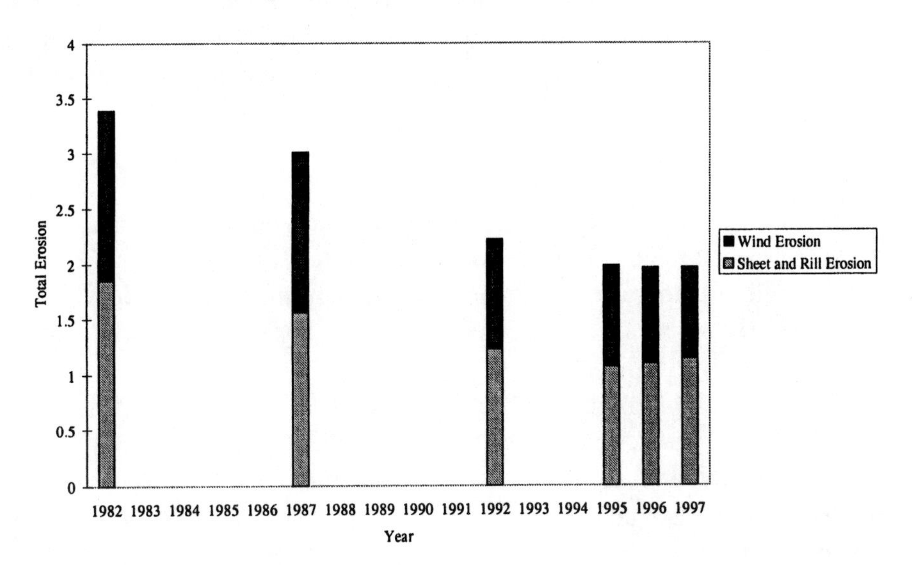

Subsequent inventories were conducted in 1987 and 1992. The 1987 NRI contained nearly 108,000 PSUs. The 1992 NRI contained more than 300,000 PSUs, including those encompassed in the 1982 NRI. Both the 1987 and 1992 NRI were designed to be compatible with the 1982 NRI in terms of data definitions, sampling techniques, etc. The 1992 NRI was different than the previous (1982 and 1987) inventories in that remote sensing techniques, particularly photo-interpretation, were used to gather much of the data instead of field visits. Special National Resources Inventories were conducted in 1995,

27 A PSU is the first stage sampling unit and is an area of land. In most parts of the country, the PSU was a square, one-half mile on a side, containing 160 acres.

28 The difficulties associated with using earlier resource inventories as the benchmark are recounted in Soil Conservation Service [1989]. The interested reader is referred to this publication.

1996, and 1997. These Special NRIs were conducted on a subsample of between 3,000 and 6,000 primary sampling units taken from the 300,000 PSUs included in the 1982 NRI and the 1992 NRI. The special inventories were conducted in accordance with procedures and standards established for the 1982, 1987, and 1992 NRIs. Consequently, they provide a scientifically credible and statistically valid basis for measuring the aggregate effects of such things as changes in erosion at the national level. Like the 1992 NRI, the 1995, 1996, and 1997 NRIs employed photo-interpretation methods. The definitions employed in the Special NRIs were consistent with those used for 1982, 1987, and 1992 thereby allowing comparisons of such things as changes in erosion rates and total erosion.

(B) RESULTS

Because the Food Security Act of 1985 singled out highly erodible land (HEL) for special treatment, it is useful to separate HEL from non-HEL (non-highly erodible land) when examining the data from the NRIs. It will help give some assessment of the effectiveness of the conservation measures that were legislatively mandated.

Total erosion on all cropland fell by 42 percent between 1982 and 1997 (Figure 6.1). In 1982, total erosion was 3,400,000,000 tons (237,000,000 tons)29 annually and in 1997, erosion was down to 2,000,000,000 tons (98,000,000 tons) annually. On just HEL cropland, aggregate erosion fell by 46.5 percent between 1982 and 1997, from 1,900,000,000 tons (132,600,000 tons) to 1,000,000,000 tons (50,700,000 tons). Between 1995 and 1997, however, there was no statistically significant[30] change in total erosion on all cropland or HEL cropland alone.

Sheet and rill erosion on all cropland fell by 38.4 percent between 1982 and 1997. In 1982, sheet and rill erosion was 1,900,000,000 tons (129,600,000 tons) annually and in 1997, it was down to 1,100,000,000 tons (45,600,000 tons) annually. Aggregate sheet and rill erosion on HEL cropland fell by 46.7 percent between 1982 and 1997, from 1,000,000,000 tons (72,300,000 tons) to 500,000,000 tons (22,000,000 tons). Between 1995 and 1997, however, there was no statistically significant change in total sheet and rill erosion on either all cropland or on just HEL cropland.

---

[29] The standard error of the estimate here and elsewhere is in parentheses.

[30] At the 95 percent level.

Wind erosion on all cropland fell by 46.3 percent between 1982 and 1997. In 1982, wind erosion was 1,500,000,000 tons (165,100,000 tons) annually and in 1997, wind was about 1,100,000,000 tons (91,500,000 tons) annually. On just HEL cropland, aggregate wind erosion fell by 46.3 percent between 1982 and 1997, from 900,000,000 tons (94,600,000 tons) to 500,000,000 tons (36,900,000 tons). Between 1995 and 1997, however, there was no statistically significant change in total wind erosion.

Erosion on all cropland declined from a rate of 8.0 tons per acre per year (0.18 tons per acre per year) in 1982 to 5.2 tons per acre per year (0.12 tons per acre per year) in 1997 (Figure 6.2). Sheet and rill erosion fell from 4.4 tons per acre per year (0.07 tons per acre per year) to 3.0 tons per acre per year (0.04 tons per acre per year) - a 31.8 percent decline - while wind erosion was reduced from 3.6 tons per acre per year (0.11 tons per acre per year) to 2.2 tons per acre per year (0.08 tons per acre per year) - a 38.9 percent fall. Between 1995 and 1997, sheet and rill erosion appeared to increase slightly, from 2.8 tons per acre per year (0.07) to 3.0 tons per acre per year (0.05 tons per acre per year) - a 7.1 percent increase - while wind erosion declined from 2.4 tons per acre per year (0.13 tons per acre per year) to 2.2 tons per acre per year (0.11 tons per acre per year) - a 8.3 percent reduction. A closer examination shows that nether of these changes is statistically significant.

Figure 6.2. Erosion Rate on Total Cropland (tons per acre per year)

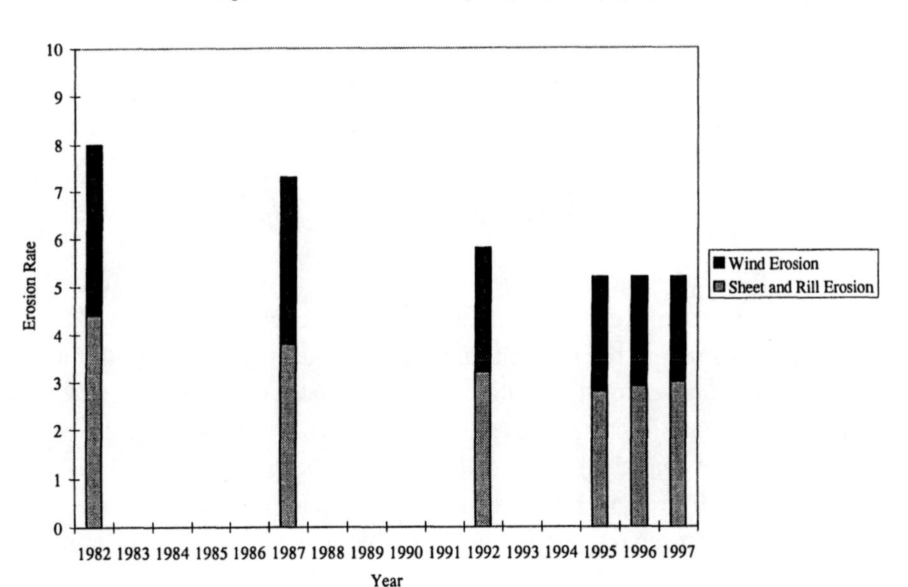

Erosion on HEL cropland between 1982 and 1997 fell from 15.1 tons per acre per year (0.19 tons per acre per year) to 9.3 tons per acre per year (0.16 tons per acre per year). Sheet and rill erosion fell from 8.2 tons per acre per year (0.07 tons per acre per year) to 5.1 tons per acre per year (0.06 tons per acre per year) - a 37.8 percent decline - while wind erosion was reduced from 6.9 tons per acre per year (0.12 tons per acre per year) to 4.2 tons per acre per year (0.10 tons per acre per year) - a 39.1 percent fall. Between 1995 and 1997, sheet and rill erosion apparently increased slightly, from 4.9 tons per acre per year (0.19 tons per acre per year) to 5.1 tons per acre per year (0.06 tons per acre per year) - a 4.1 percent increase - while wind erosion seemed to decline from 4.7 tons per acre per year (0.14) to 4.2 tons per acre per year (0.10 tons per acre per year) - a 10.6 percent reduction. These changes, however, are not statistically significant.

Erosion on non-HEL cropland between 1982 and 1997 declined from 5.0 tons per acre per year to 3.5 tons per acre per year. Sheet and rill erosion fell from 2.8 tons per acre per year (0.07 tons per acre per year) to 2.1 tons per acre per year (0.06 tons per acre per year) - a 25.0 percent decline - while wind erosion was reduced from 2.2 tons per acre per year (0.0.12 tons per acre per year) to 1.4 tons per acre per year (0.10 tons per acre per year) - a 36.4 percent fall. Between 1995 and 1997, sheet and rill erosion ostensibly increased slightly, from 2.0 tons per acre per year (0.05 tons per acre per year) to 2.1 tons per acre per year (0.04 tons per acre per year) - a 5.0 percent increase - while wind erosion declined from 1.5 tons per acre per year (0.09 tons per acre per year) to 1.4 tons per acre per year (0.09 tons per acre per year) - a 6.7 percent reduction. Neither of these changes, however, is statistically significant.

IMPLICATIONS

The NRI results are quite revealing with regard to changes in soil erosion in the United States. Between 1982 and 1995, there was a significant drop in both total erosion and the erosion rate on cropland. Additionally, the decline was most pronounced on HEL, the target of legislative efforts to reduce erosion and its offsite impacts. Since 1995, however, there has been no statistically identifiable change in either soil erosion or the erosion rate.

Even though erosion has been reduced, it continues to be a significant problem. Consider the magnitude of this problem. To do so requires a measure that quantifies the offsite social benefits of a reduction in soil erosion. This requires estimates of the offsite damages associated with erosion.

Ribaudo [1989] developed comprehensive estimates of the offsite damages associated with sheet and rill erosion.[31] The approach has been applied in a number of settings (e.g., Magleby et al. [1995] and Ribaudo et al. [1990]). The estimates take into account damage to water uses such as recreation, water storage facilities, commercial fishing, navigation, water storage, drinking water supplies, industrial water supplies, and irrigation. The estimates are compiled from an eclectic assortment of studies. These estimates will be used here with the assumption that any reduction in offsite damages translates into a comparable increase in social benefits.

Huszar and Piper [1986] have derived estimates of the offsite damages due to wind erosion. There is considerable uncertainty, however, in quantifying the damages due to wind erosion. This uncertainty is a function of the poor understanding of households' response to blowing soil and how damages vary with population density. Absent any better alternative, however, these estimates will be used with the assumption that any reduction in offsite damages will lead to a comparable increase in social benefits.

Combining the estimates of sheet and rill and wind erosion damages with data from the NRI on reductions in total erosion between 1982 and 1997, it is possible to estimate the social benefits associated with the realized reduction in erosion between 1982 and 1997. The best (most likely realized) estimate for the social benefits of a reduction in sheet and rill soil erosion is $1,110,000,000 annually. The range is between $619,320,000 and $2,120,000,000. For wind erosion, the best estimate is $1,020,000,000 annually with a range of between $737,000,000 and $2,090,000,000.

These realized social benefits are only part of the story. The unrealized social benefits (social costs) associated with continued erosion are also important especially since there has been no identifiable change in soil erosion since 1995. These unrealized benefits will continue to accrue. Based on 1997 erosion rates, the best estimate of the social costs associated with continued sheet and rill erosion is $17,810,000,000 annually with a range of between $9,940,000,000 and $33,970,000,000. The best estimate of the social costs of

[31] Ribaudo and Hellerstein [1992] discuss the methodological issues associated with estimating the water quality benefits associated with reduced sheet and rill erosion.

continued wind erosion is $11,850,000,000 annually with a range of between $8,560,000,000 and $24,240,000,000.

The social costs of continuing erosion are quite substantial, equivalent to 0.4 percent of gross domestic product. Moreover, what is clear is that programs currently in place provide insufficient incentives for a substantial increase in the number of farmers who adopt conservation practices such as conservation tillage, contour farming, filter strips, grassed waterways, terracing, polyacrylamide, and grasses and legumes in rotation. These practices can reduce soil erosion and the transport of sediments to off-farm water bodies. There are a number of policy options available, however, that can be used to induce farmers to change their production practices. These will be explored in the penultimate section.

Policies Designed to Affect the Adoption of Specific Production Practices

Several public policies can be used to affect farmers' choices of production practices and technologies: education and technical assistance, financial assistance, research and development, land retirement, and regulation and taxes. Each policy has implications about agricultural profits and the allocation of public funds.

(A) Education and Technical Assistance

If a preferred practice would be profitable for a farmer, but the farmer is unaware of its benefits, education efforts can lead to voluntary use of the practice. Educational activities generally take the form of demonstration projects and information campaigns in print and electronic media, newsletters, and meetings. Demonstration projects provide more direct and detailed information about farming practices and production systems, and how these systems are advantageous to the producer (Bosch et al. [1995]). Information assumes an especially significant role in the case of new or emerging technologies (Saha et al. [1994]). When adoption of a practice would lead to an increase in long run profits, but either new skills are needed or farming operations must be adapted for the practice to produce the highest net benefits, technical assistance can be provided to those who choose to adopt. Technical assistance is the direct, one-on-one contact provided by an assisting agency or private company for the purpose of providing a farmer with the planning and knowledge necessary to implement a particular practice on the individual farm. Requirements for successful implementation vary between individual farms because of resource conditions, operation structure, and owner/operator managerial skill. Testing a practice on part of the farm enhances its potential for adoption (Office of Technology Assessment [1990] and Nowak and O'Keefe [1995]). Technical assistance is often critical, especially for practices that require a greater level of management than the farmer currently uses (Dobbs et al. [1995]). Both education and technical assistance can be provided by either public or private sources, and both will induce adoption by farmers for whom the practice would be more profitable than the one they had been using.

(B) FINANCIAL ASSISTANCE

Financial assistance can be offered to overcome either short run or long run impediments or barriers to adoption. If the practice would be profitable once installed but involves initial investment or transition and adjustment expenses, a single cost-share payment can be used to encourage the switch to the preferred practice. Transition and adjustment costs include lost production, increased risk, or increased management costs due to learning how to use the new production practice efficiently. Financial and organizational characteristics of the whole operation also may be a hindrance to adoption (Office of Technology Assessment [1990] and Nowak [1991]). When the practice would not be more profitable to the farmer than the current practice but the environmental or other off-farm benefits are substantial, public funds could be allocated on an ongoing basis to defray the loss in profits to the farmer. Another form of financial incentive could be the granting of a tax credit for investment in a particular practice. From a public perspective, the optimal financial assistance rates are those that induce the adoption of desired practices at the least cost. Efficient rates would have to be set individually since farm and farmer characteristics vary widely (Caswell and Shoemaker [1993]). Therefore, for ease of implementation, most large financial assistance programs specify a uniform subsidy rate across resource conditions. Uniform rates, however, invariably introduce production distortions. Because resource and production characteristics can vary widely, different farms may need different sets of practices to achieve the same environmental goal. A production system that is appropriate for one farm may be inappropriate for another. The effectiveness of a conservation system in controlling erosion depends on several factors, including the frequency, timing, and/or severity of wind and precipitation, the exposure of land forms to weather, the ability of exposed soil to withstand erosive forces, the plant material available to shelter soils, and the propensity of production practices to reduce or extenuate erosive forces. An efficient financial assistance program would have a list of eligible management practices that included all alternatives appropriate for each farm. Cost-share and incentive payment policies are based on the fact that targeted farmers would not voluntarily adopt the preferred practice but the public interest calls for the practices to be used more widely. Financial assistance is not a substitute for education and technical assistance. Even with financial assistance, a farmer will not adopt a technology if he or she is unfamiliar with it.

(C) Research and Development (R&D)

Research and development policies can be used to enhance the benefits of a given production practice. The objective of the research would be to either improve the performance or to reduce the costs of the practice. Data gathering and analysis, as well as monitoring also contribute to R&D by providing information necessary to assess the determinants of adoption and the effectiveness of practices in achieving public goals. In addition, R&D funds could be allocated to ensure that the practice is adaptable for more circumstances. R&D is a long run policy strategy with an uncertain probability of success, but it may also reap the greatest gains in encouraging the voluntary adoption of a preferred technology because it can increase the profitability of the practice for a wider range of potential adopters.

(D) Land Retirement

The policy that has the largest impact on farmers' choice of practices or technologies is land retirement. The underlying premise is that large public benefits can be gained by radically changing agricultural practices on particular parcels of land and that changes in individual practices would not provide sufficient social benefits. For an individual to voluntarily agree to put the land in conserving uses, he or she would expect compensation in an amount at least as great as the lost profits from production. The payment mechanisms that can be used to implement land retirement strategies are lump sum payments or annual "rental fees." The former are often referred to as easements whereby the farmer's right to engage in nonconserving uses is purchased by the public sector for a specific period. Payment to an individual to retire land would result in a voluntary change in practices.

(E) REGULATION, TAXES, AND INCENTIVES

If voluntary measures prove insufficient to produce the changes in practices necessary to achieve public goals, regulation is a policy that can be used. The use of certain practices could be prohibited, taxed, or made a basis for withholding other benefits. Preferred practices could be required or tax incentives offered to promote their use thereby offsetting some of the cost of the requisite new equipment. Point sources of pollution have been subject to command-and-control policies for many years. There are recognized inefficiencies associated with technology-based regulations because the least-cost technology combination to meet an environmental goal for an individual may not be permitted.[32] It has been assumed that such loss in efficiency is made up for by ease of implementation.

CONCLUSION

Soil erosion has both on-farm and off-farm impacts. Reduction of soil depth can impair the land's productivity, and the transport of sediments can degrade streams, lakes, and estuaries.

Soil conservation policies have existed in the United States for over 60 years. Initially, these policies focused on the on-farm benefits of keeping soil on the land and increasing net farm income. Beginning in the 1980s, however, policy goals increasingly included reductions in off-site impacts of erosion. The Food Security Act of 1985 was the first major legislation explicitly to tie eligibility to receive agricultural program payments to conservation performance. The Federal Agriculture Improvement and Reform Act (FAIR) of 1996 modifies the conservation compliance provisions by providing farmers with greater flexibility in developing and implementing conservation plans. Noncompliance for those on HEL can result in loss of eligibility for many payments including production flexibility contract payments.

As a consequence of conservation efforts, total soil erosion between 1982 and 1997 was reduced by 42 percent and the erosion rate fell from 8.0 tons per

[32] Economic theory shows that the efficient solution (i.e., least cost for society to achieve a particular level of environmental quality) is when the marginal cost of pollution reduction is the same for all producers (Baumol and Oates [1988]). Each individual could have different combinations of practices and technologies. To implement such a policy, however, would have an extremely high cost for an industry as large and diverse as agriculture (Hefferman [1984]).

acre per year in 1982 to 5.2 tons per acre per year in 1997. Still, soil erosion is imposing substantial social costs. In 1997 these costs are estimated to have been about $29,700,000,000.

To further reduce soil erosion and thereby mitigate its social costs, there are a number of policy options available to induce farmers to adopt conservation practices including education and technical assistance, financial assistance, research and development, land retirement, and regulation and taxes.

REFERENCES

Barker, J., G. Baumgardner, D. Turner, and J. Lee, "Carbon Dynamics of the Conservation and Wetland Reserve Programs," *Journal of Soil and Water Conservation*, Vol. 51 (1996), pp. 340-346.

Baumol, W., and W. Oates, *The Theory of Environmental Policy*, Second Edition, Cambridge University Press, Cambridge, 1988.

Bosch, D., Z. Cook, and K. Fuglie, "Voluntary Versus Mandatory Agricultural Policies to Protect Water Quality: Adoption of Nitrogen Testing in Nebraska," *Review of Agricultural Economics*, Vol. 17 (1995), pp. 13-24.

Caswell, M., and R. Shoemaker, *Adoption of Pest Management Strategies Under Varying Environmental Conditions*, Technical Bulletin 1827, Economic Research Service, U.S. Department of Agriculture, Washington, DC, December 1993.

Crosswhite, W., and C. Sandretto, "Trends in Resource Protection Policies in Agriculture," *Agricultural Resources: Cropland, Water, and Conservation Situation and Outlook Report*, Economic Research Service, U.S. Department of Agriculture, Washington, DC, 1991.

Dobbs, T., J. Bischoff, L. Henning, and B. Pflueger, "Case Study of the Potential Economic and Environmental Effects of the Water Quality Incentives Program and the Integrated Crop Management Program: Preliminary Findings," Paper presented at annual meeting of the Great Plains Economics Committee, Great Plains Agricultural Council, Kansas City, MO, April 1995.

Foster, G., and S. Dabney, "Agricultural Tillage Systems: Water Erosion and Sedimentation," Farming for a Better Environment, Soil and Water Conservation Society, Ankeny, IA, 1995.

Gebhart, D., H. Johnson, H. Mayeux, and H. Polley, "The CRP Increases Soil Organic Carbon," *Journal of Soil and Water Conservation*, Vol. 49 (1994), pp. 488-492.

Hefferman, W., "Assumptions of the Adoption/Diffusion Model and Soil Conservation," in B. English, J. Maetzold, B. Holding, and E. Heady (eds.) *Agricultural Technology and Resource Conservation*, Iowa State University Press, Ames, IA, 1984.

Heimlich, R., "Soil Erosion and Conservation Policies in the United States," in N. Hanley (ed.) *Farming and the Countryside: An Economic Analysis of External Costs and Benefits*, CAB International Publishers, Miami, FL, 1991.

Holmes, B., *Legal Authorities for Federal (USDA), State, and Local Soil and Water Conservation Activities: Background for the Second RCA Appraisal*, U.S. Department of Agriculture, Washington, DC, 1987.

Huzsar, P., and S. Piper, "Estimating Offsite Costs of Wind Erosion in New Mexico," Journal of Soil and Water Conservation, Vol. 41 (1986), pp. 414-416.

Karlen, D., "Conservation Tillage Research Needs," *Journal of Soil and Water Conservation*, Vol. 45 (1990), pp. 365-369

Libby, L., "Interaction of RCA with State and Local Conservation Programs," in H. Halcrow, E. Heady, and M. Cotner (eds.), *Soil Conservation Policies Institutions, and Incentives*, Soil Conservation Society of America, Ankeny, IA, 1982.

Lindstrom, M., T. Schumacher, A. Jones, and C. Gantzer, "Productivity Index Model for Selected Soils in North Central United States," *Journal of Soil and Water Conservation*, Vol. 47 (1992), pp. 491-494.

Magleby, R., C. Sandretto, W. Crosswhite, and T. Osborn, Soil Erosion and Conservation in the United States, AIB-718, Economic Research Service, U.S. Department of Agriculture, Washington, DC, 1995.

Missouri Management Systems Evaluation Area, *A Farming Systems Water Quality Report*, Research and Education Report, Missouri MSEA, 1995.

Moldenhauer, W., W. Kemper, and G. Langdale, "Long Term Effects of Tillage and Crop Residue Management," in G. Langdale and W. Moldenhauer (eds.), Crop Residue Management to Reduce Erosion and Improve Soil Quality - Southeast, CR-39, Agricultural Research Service, U.S. Department of Agriculture, Washington, DC, 1994.

National Research Council, Soil and Water Quality, National Academy Press, Washington, DC, 1993.

Natural Resources Conservation Service, *A Geography of Hope*, U.S. Department of Agriculture, Natural Resources Conservation Service, Washington, DC, 1996.

Nelson, F., and L. Schertz, *Provisions of the Federal Agriculture Improvement and Reform Act of 1996*, Economic Research Service, U.S. Department of Agriculture, Washington, DC, 1996.

Nowak, P., "Why Farmers Adopt Production Technology," in *Crop Residue Management for Conservation*, Proceedings of a National Conference, Soil and Water Conservation Society, Ankeny, IA, 1991.

Nowak, P., and G. O'Keefe. 1995. "Farmers and Water Quality: Local Answers to Local Issues," Draft Report submitted to the U.S. Department of Agriculture, Washington, DC, September 1995.

Osborn, T., "Conservation," *Provisions of the Federal Agricultural Improvement Act of 1996*, Economic Research Service, U.S. Department of Agriculture, Washington, DC, 1996.

Piper, S., and L. Lee, Estimating the Offsite Household Damage from Wind Erosion in the Western United States, Staff Report 8926, Economic Research Service, U.S. Department of Agriculture, Washington, DC 1989.

Rasiah, R., and B. Kay, "Characterizing the Changes in Aggregate Stability Subsequent to Introduction of Forages," *Soil Science Society of America Journal*, Vol. 58 (1994), pp. 935-942.

Rasmussen, W., "History of Soil Conservation, Institutions and Incentives," Soil Conservation Policies, Institutions, and Incentives, Soil Conservation Society of America, Ankeny, IA, 1982.

Reichelderfer, K., "Land Stewards or Polluters? The Treatment of Farmers in the Evaluation of Environmental and Agricultural Policy," a paper presented at the conference Is Environmental Quality Good for Business?, Washington DC, 1990.

Ribaudo, M., *Water Quality Benefits from the Conservation Reserve Program, AER-606*, Economic Research Service, U.S. Department of Agriculture, Washington, DC, 1989.

Ribaudo, M., D. Colacicco, L. Langer, S. Piper, and G. Schaible, *Natural Resources and Users Benefit from the Conservation Reserve Program, AER-627*, Economic Research Service, U.S. Department of Agriculture, Washington, DC, 1990.

Ribaudo, M., and D. Hellerstein, *Estimating Water Quality Benefits: Theoretical and Methodological Issues, TB-1808*, Economic Research Service, U.S. Department of Agriculture, Washington, DC, 1992.

Saha, A., H.A. Love, and R. Schwart, "Adoption of Emerging Technologies Under Output Uncertainty," *American Journal of Agricultural Economics*, Vol. 76 (1994), pp. 836-846.

Schumacher, T., M. Lindstrom, M. Blecha, and R. Blevins, "Management Options for Lands Concluding their Tenure in the Conservation Reserve Program," in R. Blevins and W. Moldenhauer (eds.) *Crop Residue Management to Reduce Erosion and Improve Soil Quality - Appalachia and Northeast*, Agricultural Research Service, U.S. Department of Agriculture, Washington, 1995.

Soil Conservation Service, *Summary Report 1987 National Resources Inventory*, SB 790, Soil Conservation Service, Washington, DC, 1989.

Strohbehn, R., *An Economic Analysis of USDA Erosion Control Programs: A New Perspective, AER-560*, Economic Research Service, U.S. Department of Agriculture, Washington, DC 1986.

U.S. Congress, Office of Technology Assessment, *Beneath the Bottom Line: Agricultural Approaches to Reduce Agrichemical Contamination of Groundwater*, OTA-F-418, U.S. Government Printing Office, Washington, DC, November 1990.

U.S. Environmental Protection Agency, Office of Water, National Water Quality Inventory - 1994 Report to Congress, Office of Water, Washington, DC, 1995.

van Es, J. "Dilemmas in the Soil and Water Conservation Behavior of Farmers," in B. English, J. Maetzold, B. Holding, and E. Heady (eds.) *Agricultural Technology and Resource Conservation*, Iowa State University Press, Ames, 1984.

Zison, S., K. Haven, and W. Mills, Water Quality Assessment: A Screening Method for Nondesignated 208 Areas, Environmental Protection Agency, Athens, GA 1977.

CHAPTER 7

PERCEPTIONS VERSUS REALITY: THE CASE OF NO TILL FARMING

INTRODUCTION

No till is an agricultural production practice where the soil is left undisturbed from harvest to seeding and from seeding to harvest.[33] The only tillage is the soil disturbance in a narrow slot created by coulters, disk or runner seed furrow openers, or hoe openers attached to the planter or drill. No till planters must be able to cut residue and penetrate undisturbed soil. Weed control relies on herbicides applied preplant, pre-emerge or post-emerge. The type and timing of herbicide application depends on the weed pressures and climatic conditions. Strictly speaking, no till does not allow operations that disturb the soil other than the planting operation (MidWest Plan Service [1992]).

BENEFITS OF NO TILL

There are a number of significant economic and environmental benefits associated with the use of no till in production agriculture in the United States.

---

[33] No till is one type of conservation tillage. Conservation tillage is defined to be any tillage and planting system that maintains at least 30 percent of the soil surface covered by residue after planting to reduce soil erosion by water. Where soil erosion by wind is the primary concern, any system that maintains at least 1,000 pounds (per acre) of flat, small grain residue equivalent on the surface during the critical wind erosion period. Two key factors influencing crop residue are (1) the type of crop, which establishes the initial residue amount and determines its fragility, and (2) the type of tillage operations prior to and including planting (Conservation Technology Information Center [1998]).

(A) ECONOMIC BENEFITS OF NO TILL

With regard to the economic benefits, the use of no till may affect the cost of labor, fertilizers, pesticides, and machinery relative to conventional tillage (Zero Tillage Farmers Association [1997]).[34]

(i) Labor Use and Cost

A reduction in the intensity and number of tillage operations lowers costs for labor and machinery, especially if the machinery is used optimally (Siemans and Doster [1992]). Several studies estimate the savings in labor costs if no till is adopted. Weersink et al. [1992], for example, found that corn-soybean farmers in southern Ontario realized significant savings in labor costs with no till compared with conventional tillage systems. The omission of preplant operations alone reduces labor requirements by up to 60 percent.

Dickey et al. [1992] calculated the typical labor requirements from machinery management data for various tillage systems in Nebraska. Compared with the commonly used disk system, no till saves about 20 minutes of labor per acre.

NASS/ERS Cropping Practices Survey (U.S. Department of Agriculture [1996]) data also show that labor savings can be significant. The number of hours devoted to tillage operations (including stalk chopping and planting) is different between no till and conventional tillage, but the relative amount of time spent on tillage operations has not appreciably changed over time. In 1995, conventional tillage operations for corn took 0.38 hour per acre while no till required only 0.13 hour per acre. Soybeans and wheat took, respectively, 0.44 and 0.47 hour per acre for conventional tillage while no till operations on average took 0.10 and 0.12 hour per acre. All of these differences are statistically significant. Moreover, the relative amount of time devoted to the tillage operations for no till and conventional tillage has not changed between 1990 and 1995.

The benefit from no till of reduced labor needs is greater than just the labor cost savings per acre. There is the associated opportunity cost of the labor and time saved. That is, if less hired labor is needed, there will be direct savings. Saving the farmer's or other family labor may permit them to engage in

[34] Conventional tillage includes tillage types that leave less than 15 percent residue cover after planting, or less than 500 pounds per acre of small grain residue equivalent through the critical wind erosion period. This generally includes plowing or other intensive tillage. Weed control is accomplished with herbicides and/or cultivation.

off-farm activities. Lower labor requirements for tillage lead to additional returns from the expansion of existing enterprises or allow time for new activities to improve profitability for the whole farm operation.

(ii) Fertilizer Use and Cost

The determination of fertilizer needs to attain optimum yields depends on an accurate assessment of the soil's available nutrients in relation to the needs of the crop. This assessment will include site-specific factors such as soil characteristics, cropping patterns, and climatic conditions in addition to the tillage practice. No till, however, requires improved fertilizer management (Halvorson [1994] and Rehm [1995]). In some instances, increased application of specific nutrients may be necessary and specialized equipment required for proper fertilizer placement, thereby contributing to higher costs.

The results of field experiments indicate that an increase in the amount of crop residue cover on the soil surface tends to keep soils cooler, wetter, less aerated, and denser (Mengel et al. [1992]). These characteristics and beneficial impacts from increased organic matter, improved moisture retention and permeability, and reduced nutrient losses from erosion are associated with no till and can affect the ability of crops to utilize nutrients. With higher levels of crop residue, proper timing and placement of nutrient applications are critical to enhance fertilizer efficiency to achieve optimal yield at lowest cost.

(iii) Pesticide Use and Cost

Weed control problems vary among tillage systems because the nature of the weed population changes. Tillage prepares a seedbed not only for the crop but for weed seeds as well (Monson and Wollenhaupt [1995]). Different weed species occur as tillage is reduced, requiring different control programs. Effective weed control with herbicides depends on spraying at the right stage of plant growth, plant stress, weather conditions, and so on. Weed growth and development, as well as appropriate management strategies, vary with location. A weed management program must be site specific and circumstance specific and will be different between no till and conventional tillage (Martin [1992]).

The reservoir of dormant weed seeds resident in the soil will not be transferred to the germination zone near the soil surface by tillage. Consequently, as annual weeds are controlled, the overall weed problem may decrease after a few years when fields are converted to no till and if effective weed control is practiced (Fawcett [1987]).

(iv) Machinery Use and Costs

Machinery-related costs typically range from $50 per acre per year to $70 per acre per year (Siemans and Doster [1992]) and overshadow all other cost categories except land. As with the choice of other inputs, a farmer endeavors to perform the requisite field operations with the optimum or least cost machinery inventory. Because the optimum machinery inventory differs across tillage systems, direct comparison of machinery costs that can perform the desired field operations is difficult. As the size (width) of a machinery set increases, machinery productivity increases. But the annual machinery costs, fixed and variable, also increase with increasing machinery size. For no till, fewer implements and field operations are used. If no till is used on only part of the cropland, however, implements and tractors will need to be available for other portions, so cost comparisons will be more difficult. Thus, using a drill or narrow-row planter for soybeans is an option for most tillage systems. Owning a drill, however, for soybeans and a planter for corn often increases the machinery inventory and costs for a corn-soybean farm.

Additionally, a farmer who decides to convert from conventional tillage to no till exclusively must consider how to value his or her existing conventional tillage equipment. This equipment might not be fully depreciated and farmers often have limited alternatives on how it might otherwise be used (Bates et al. [1979] and John Deere and Co. [1980]). A complete assessment of the production costs associated with no till versus conventional tillage must make some provision for the opportunity cost of the conventional tillage equipment. There are no data available, however, to permit a comparison of machinery costs between tillage systems.

Machinery operating costs may not be lower for no till. Lower fuel and maintenance costs associated with no till may be overshadowed by the higher cost of new no till implements. Lower fuel costs are a consequence of the fewer trips across a field with no till. Maintenance costs will be lower because the equipment will be used less (Hunt [1984]).

(B) Environmental Benefits of No Till

No till serves to reduce soil erosion. For example, a thirty percent crop residue cover associated with the use of no till, results in a reduction in soil loss by 65 percent relative to conventional tillage while a 60 residue cover results in an 88 percent reduction in soil loss (University of Illinois Agricultural Extension Service [1997]).

A reduction in soil erosion associated with no till will mitigate many offsite erosion-related problems including water use impairment. The most widespread offsite erosion-related problem is impairment of water resource use (National Research Council [1993]). The Environmental Protection Agency has identified siltation associated with erosion in rivers and lakes as the second leading cause of water quality impairment, and agricultural production is identified as the leading cause of water quality impairment (U.S. Environmental Protection Agency [1995]).

Three related causes of water use impairment are sedimentation, eutrophication, and pesticide contamination. When soil particles and agricultural chemicals wash off a field, they may be carried in runoff until discharged into a water body or stream. Not all agricultural constituents that are transported from a field reach water systems, but a significant portion does, especially dissolved chemicals and the more chemically active, finer soil particles. Once agricultural pollutants enter a water system, they lower water quality and can impose economic losses on water users. These offsite impacts can be substantial. The offsite impacts of erosion are potentially greater than the onsite productivity effects in the aggregate (Foster and Dabney [1995]). Therefore, society may have a larger incentive for reducing erosion than farmers have.

If the runoff reaches the water body or stream, soil particles can be suspended in the water, or settle out as sediment, depending on the velocity of the waterflow and the size of the soil particles. In each case, water use can be affected.

Suspended soil particles affect the biologic nature of water systems by reducing the transmission of sunlight, raising surface water temperatures, and affecting the respiration and digestion of aquatic life. The effects on aquatic life, and the reduction in aesthetic quality of recreation sites, can reduce the value of water for recreation uses. Suspended soil particles impose costs on water treatment facilities which must filter out the particles. Suspended soil particles can also damage moving parts in pumps and turbines.

Even when soil particles settle on the bottom of a river or lake, they can cause serious problems for aquatic life by covering food sources, hiding places, and nesting sites. Sedimentation can clog navigation and water conveyance systems like roadside ditches, reduce reservoir capacity, and damage recreation sites. In streambeds, sedimentation can lead to an increase in the frequency and severity of flooding by reducing channel capacity.

The nutrients and pesticides attached to soil particles, or dissolved in runoff, affect water quality in ways that can affect the suitability of water for many uses (Baker [1987]). The most far-reaching impact is eutrophication, abundant growth of algae and rooted vegetation caused by excessive nutrient runoff. As algae dies and decays, it uses oxygen from the surrounding water, lowering the dissolved oxygen levels and altering the size and composition of commercial and recreational sport fisheries. Rooted plants can become a nuisance around marinas and shorelines. Floating algae blooms can restrict light penetration to surface waters and can affect the health, safety, and enjoyment of people using water for recreation. Floating algae can clog intake pipes and filtration systems, increasing the cost of water treatment.

Pesticides, which include herbicides, insecticides, and fungicides, create a broad array of impacts. Most notable are effects on aquatic life. Very high concentrations will kill organisms outright. Lower concentrations, more commonly observed, can produce a variety of sublethal effects such as to lower resistance of fish, which makes them susceptible to other stresses (Glotfelty [1987]). Herbicides can hinder photosynthesis in aquatic plants (Schepers [1987]). Pesticides can damage commercial and sport fisheries and make fish dangerous to eat (Herndon [1987]).

Several studies are available that evaluate the improvements in water quality associated with the use of no till. A few of these are reviewed. Richards and Baker [1998] report on the effort to reduce the eutrophication in Lake Erie that began in the early 1970s. A monitoring station was set up at Bowling Green, Ohio on the Maumee River which feeds into Lake Erie. Agriculture is the dominant land use in the Maumee River basin. Major crops are corn, soybeans, and wheat. Between 1975 and 1995, implementation of no till increased from less than 5 percent to more than 50 percent of planted acreage. Fertilizer (nitrogen and phosphorous) application rates also changed over the period so it is not possible to quantify precisely the contribution of no till to the water quality improvements. Nevertheless, water quality changes over the study period were evaluated by conducting trend studies of concentrations and loads. The adoption of no till in conjunction with the reduced fertilizer application rates led to a reduction in total phosphorous loadings of 24 percent, a reduction in suspended sediments of 19 percent, and a reduction in total Kjeldahl nitrogen of 10 percent.

Fawcett et al. [1994][35] survey the effects of various best management practices, including no till, on pesticide runoff into surface water and leaching into groundwater. They conclude that no till systems provide a reduction in runoff losses for active pesticide ingredients studied. Average[36] herbicide runoff in no till systems, for example, was 30 percent of the conventional tillage runoff. Additionally, ridge tillage and chisel plow practices are less effective than no till in reducing soil erosion on HEL but are relatively good production practices on less erodible fields. For the various no till practices relative to conventional tillage, herbicide runoff was 70 less for no till while it was 42 percent less for ridge tillage. With regard to leaching, however, no till does not fare as well. Increases in infiltration accompanying the use of no till may result in a greater threat to groundwater from pesticides or nitrate. Preferential flow of water through macropores, which may be more prevalent with no till, can allow water and dissolved solutes or suspended sediment to bypass upper layers of soil. This may transfer pesticides to shallow groundwater or to depths in the soil where biological degradation is slower. It is important, however, to keep this in perspective. Even though there in an increase in the potential leaching risks of certain pesticides associated with no till, the relative concentrations of pesticides found in surface water are typically greater than concentrations in groundwater.

Wind erosion produces offsite impacts that can be as dramatic as the Dust Bowl of the 1930s. It has not, however, received the attention given to the more widespread water erosion impacts. Damage can include higher maintenance of buildings and landscaping, pitting of automobile finishes and glass, greater wear on machinery parts, increased soiling and deterioration of retail inventories, costs of removing blown sand and dust from roads and ditches, and increased respiratory and eye disorders. Offsite damages from wind erosion depend on the extent and location of population centers relative to prevailing winds and wind erosion sources (Piper and Lee [1989]). Consequently, damage estimates for one area cannot readily be extrapolated to other areas, nor can the impact of wind erosion from cropland or other agricultural land be differentiated from wind erosion originating on nonagricultural land.

---

[35] There is also a companion piece that repeats the conclusions of this study. See Ciba Geigy Corporation [1992].

[36] The conditions under which the field experiments are performed and from which averages are computed are critical. Basta et al. [1997] explore this issue.

PERCEPTIONS VERSUS REALITY

(A) OVERVIEW

From the foregoing, it is clear that no till is associated with a number of economic and environmental benefits. It is this realization that has been used to promote the adoption of no till in the United States (Zero Tillage Farmers Association [1997]). Between 1989 and 1997, the use of no till has increased from to 5.1 percent to 15.6 percent of total planted acreage in production agriculture in the United States (Conservation Technology Information Center [1997]).

One of the current problems with promoting the use of no till is the disparity in farmers' perception of precisely what constitutes no till. The commonly used definition of no till was noted above. There are others types of conservation tillage, however, including mulch till and ridge till.[37] The economic and environmental benefits accruing to no till do not, for the most part, carry over to other types of conservation tillage. If they do, however, the realized benefits are substantially less (Dickey et al. [1992], Fawcett et al. [1994], and University of Illinois Agricultural Extension Service [1997]).

The extent of this misperception in what constitutes no till is the subject of what follows. It is based on an analysis of survey data.

(B) DATA

In 1996, the Agricultural Resources Management Study (ARMS) survey was conducted by the U.S. Department of Agriculture. Annual data were collected on fertilizer and pesticide use, tillage systems employed, cropping sequences, whether the cropland is designated as highly erodible, and information on the use of other inputs and production practices. The survey covered corn, cotton, soybeans, wheat (winter, spring and durum), and potatoes. Since, however, no till is not a viable production practice for potatoes and it is not yet extensively used on cotton, these crops will be omitted from consideration. Only selected States were surveyed, but about 80 percent of the total planted

[37] With mulch tillage, the soil is disturbed prior to planting. Tillage tools such as chisels, field cultivators, disks, sweeps, or blades are used. Weed control is accomplished with herbicides and/or cultivation. With ridge tillage the soil is left undisturbed from harvest to planting except for nutrient injection. Planting is completed in a seedbed prepared on ridges with sweeps, disk openers, coulters, or row cleaners. Residue is left on the surface between ridges. Weed control is accomplished with herbicides and/or cultivation. Ridges are rebuilt during cultivation.

acreage for the respective crops was covered. Five tillage systems, including conventional tillage with moldboard plow, conventional tillage without moldboard plow, mulch tillage, ridge tillage, and no till, are defined based on the use of specific tillage implements and their residue incorporation rates (Bull [1993]). For the purpose of classification of the survey results, no till is defined as a system that has no till operation before planting. This does allow field passes of implements, such as stalk choppers, which do not incorporate any residue. Respondents (farmers) were also asked on the survey if they used no till.

(C) SURVEY RESULTS

Focusing on the five survey crops - corn, soybeans, winter wheat, spring wheat, and durum wheat - a comparison of farmers' perception that they used no till and whether they de facto used no till is presented in Table 7.1. Remembering that the values presented in the table are statistics, for the production of soybeans, winter wheat, spring wheat, and durum wheat, there is no statistically significant difference at the 95 percent level between farmers' perception that they used no till and their actual use of no till. Thus, the economic and environmental benefits that result from the use of no till in the production of these commodities can be fully claimed by farmers producing soybeans, winter wheat, spring wheat, and durum wheat.

The use of no till in the production of corn is different. There is a statistically significant difference at the 95 percent level in farmers' perception and reality. Nearly 18 percent of corn farmers perceive that they used no till while, in fact, only slightly more than 12 percent did use no till. A disaggregated analysis of the data show that of the nearly 18 percent of farmers who perceive they used no till, 3.4 (0.5)[38] percent actually used mulch tillage and 2.4 (0.4) percent used ridge tillage. None of these farmers, however, used conventional tillage. While the percentage of farmers using a tillage system different than no till while believing they were using no till is relatively small, the fact that 79,507,000 acres of corn were planted in 1996 translates into a substantial number of acres on which farmers perceptions with regard to tillage system used were inconsistent with the actual tillage system used (U.S. Department of Agriculture [1998]). This has important implications because the production of corn in the United States is concentrated in regions that are most suscepti-

[38] The standard error of the estimate is in parentheses.

ble to sheet and rill erosion (Conservation Technology Information Center [1997]). It is these regions where the environmental benefits associated with the use of no till are greatest (Uri [1998]).

Table 7.1. The Proportional Use of No Till

Crop	Sample Size	Indicated	Actual
Corn	3403	0.1799	0.1220
		(0.0065)[1]	(0.0056)
Soybeans	2688	0.3146	0.3227
		(0.0089)	(0.0090)
Winter Wheat	898	0.0294	0.0335
		(0.0056)	(0.0060)
Spring Wheat	252	0.0230	0.0380
		(0.0094)	(0.0121)
Durum Wheat	98	0.0515	0.0655
		(0.0224)	(0.0251)

[1] Standard errors of the estimates in parentheses.

It is interesting to speculate on why the perceived use of no till differs from its actual use for corn but not for the other commodities. The survey does not provide a definition of what constitutes no till. Thus, corn farmers might have mistaken no till for other types of conservation tillage if the soil was not tilled immediately prior to planting. Alternatively, there is a possibility that farmers were actually using no till but the classification was inaccurate. There is no reason, however, to believe that this is the situation.

CONCLUSION

A number of economic and environmental benefits are associated with the use of no till in production agriculture in the United States. There are lower labor, energy, and machinery costs associated with no till farming relative to conventional tillage systems and other types of conservation tillage. The reduced erosion associated with no till also leads to a number of environmental benefits including a reduction in water quality impairment.

In order to fully account for the benefits associated with no till, it is important that farmers' perception of what constitutes no till and their actual use of no till be consistent. An analysis of Agricultural Resource Management Study survey data for 1996 shows that for soybeans, winter wheat, spring wheat, and durum wheat, farmers' perceptions are consistent with reality. In the case of corn, however, nearly 18 percent of corn farmers believe they are using no till while in actuality, only slightly more than 12 percent are using this tillage system.

REFERENCES

Alt, K., T. Osborn, and D. Colocicco, *Soil Erosion: What Effect on Agricultural Productivity, AIB-556*, Economic Research Service, U.S. Department of Agriculture, Washington, DC, 1989.

Baker, D., "Overview of Rural Nonpoint Pollution in the Lake Erie Basin," *Effects of Conservation Tillage on Groundwater Quality*, Lewis Publishers, Chelsea, MI, 1987.

Basta, N., R. Huhnke, and J. Stiegler, "Atrazine Runoff from Conservation Tillage Systems," *Journal of Soil and Water Conservation*, Vol. 52 (1997), pp. 44-48.

Bates, J., A. Rayner and P. Custance, "Inflation and Farm Tractor Replacement in the U.S.: A Simulation Model," *American Journal of Agricultural Economics*, Vol. 61 (1979), pp. 331-334.

Bull, L., *Residue and Tillage Systems for Field Crops, AGES 9310*, Economic Research Service, U.S. Department of Agriculture, Washington, DC, 1993.

Ciba-Geigy Corporation, *Best Management Practices to Reduce Runoff of Pesticides into Surface Water: A Review and Analysis of Supporting Research, TR-9-92*, Agricultural Group, Ciba-Geigy Corporation, Greensboro, NC, 1992.

Conservation Technology Information Center, *National Crop Residue Management Survey - 1997 Survey Results*, West Lafayette, IN, 1997.

John Deere and Company, *Fundamentals of Machinery Operation-Machine Management*, Moline, IL, 1980.

Dickey, E., P. Jasa, and D. Shelton, "Conservation Tillage Systems," *Conservation Tillage Systems and Management*, MidWest Plan Service, Iowa State University, Ames, 1992.

Fawcett, R., "Overview of Pest Management Systems," in T. Logan, J. Davidson, J. Baker, and M. Overcash (eds.), *Effects of Conservation Tillage on Groundwater Quality: Nitrates and Pesticides*, Lewis Publishers, Chelsea, MI, 1987, pp. 90-112.

Fawcett, R., B. Christensen, and D. Tierney, "The Impact of Conservation Tillage on Pesticide Runoff into Surface Water: A Review and Analysis," *Journal of Soil and Water Conservation*, Vol. 49 (1994), pp. 126-135.

Foster, G., and S. Dabney, "Agricultural Tillage Systems: Water Erosion and Sedimentation," *Farming for a Better Environment, Soil and Water Conservation Society*, Ankeny, IA, 1995.

Glotfelty, D., "The Effects of Conservation Tillage Practices on Pesticide Volatilization and Degradation," *Effects of Conservation Tillage on Groundwater Quality*, Lewis Publishers, Chelsea, MI, 1987.

Halvorson, A., "Fertilizer Management," *Crop Residue Management to Reduce Soil Erosion and Improve Soil Quality: Northern Great Plains*, Agricultural Research Service, U.S. Department of Agriculture, Washington, DC, 1994.

Herndon, L., "Conservation Systems and Their Role in Sustaining America's Soil, Water, and Related Natural Resources," *Optimum Erosion Control at Least Cost*, American Society of Agricultural Engineers, St. Joseph, MI, 1987.

Hunt, D., "Farm Machinery Technology: Performance in the Past, Promise for the Future," in B. English (ed.), *Future Agricultural Technology and Resource Conservation*, Iowa State University Press, Ames, 1984.

Magleby, R., C. Sandretto, W. Crosswhite, and T. Osborn, *Soil Erosion and Conservation in the United States, AIB-718*, Economic Research Service, U.S. Department of Agriculture, Washington, DC, 1995.

Martin, A., "Weed Control," *Conservation Tillage Systems and Management*, MidWest Plan Service, Iowa State University, Ames, 1992.

Mengel, D., J. Moncrief, and E. Schulte, "Fertilizer Management," *Conservation Tillage Systems and Management*, MidWest Plan Service, Iowa State University, Ames, 1992.

MidWest Plan Service, *Conservation Tillage Systems and Management*, MidWest Plan Service, Iowa State University, Ames, 1992.

Monson, M., and N. Wollenhaupt, "Residue Management: Does It Pay?" *Crop Residue Management to Reduce Soil Erosion and Improve Soil Quality: North Central Region*, Agricultural Research Service, U.S. Department of Agriculture, Washington, DC, 1995.

National Research Council, *Soil and Water Quality*, National Academy Press, Washington, DC, 1993.

Piper, S., and L. Lee, *Estimating the Offsite Household Damage from Wind Erosion in the Western United States, Staff Report 8926*, Economic Research Service, U.S. Department of Agriculture, Washington, DC 1989.

Rehm, G., "Tillage Systems and Fertilizer Management," *Farming for a Better Environment, Soil and Water Conservation Society*, Ankeny, IA, 1995.

Richards, R., and D. Baker, *Twenty Years of Change: The Lake Erie Agricultural Systems for Environmental Change (LEASEQ) Project*, Water Quality Laboratory, Heidelberg College, Tiffin, OH, 1998.

Schepers, J., "Effects of Conservation Tillage on Processes Affecting Nitrogen Management," *Effects of Conservation Tillage on Groundwater Quality*, Lewis Publishers, Chelsea, MI, 1987.

Siemans, J., "Soil Management and Tillage Systems," *Illinois Agronomy Handbook*, University of Illinois Press, Urbana, 1997

Siemans, J., and D. Doster, "Costs and Returns," *Conservation Tillage Systems and Management*, MidWest Plan Service, Iowa State University, Ames, 1992.

University of Illinois Agricultural Extension Service, *No-Till's Benefits and Costs*, Urbana, IL, 1997.

United States Department of Agriculture, *Cropping Practices Survey - 1995*, Economic Research Service, Washington, DC, 1995.

United States Department of Agriculture, Statistical Highlights of U.S. Agriculture, National Agricultural Statistics Service, Washington, DC, 1998.

United States Environmental Protection Agency, Office of Water, *National Water Quality Inventory - 1994 Report to Congress*, Office of Water, Washington, DC, 1995.

Uri, N., *Conservation Tillage in U.S. Agriculture: Environmental, Economic, and Policy Issues*, The Haworth Press, Inc., Binghamton, NY, 1999.

A. Weersink, G. Fox, G. Sarwar, S. Duff, and B. Deen, "Comparative Economics of Alternative Agricultural Production Systems: A Review," *Northeastern Journal of Agricultural and Resource Economics*, Vol. 20 (1992), pp. 124-142.

Zero Tillage Farmers Association, *Advancing the Art*, Manitoba, ND, 1997.

Zison, S., K. Haven, and W. Mills, *Water Quality Assessment: A Screening Method for Nondesignated 208 Areas*, Environmental Protection Agency, Athens, GA 1977.

CHAPTER 8

THE USE OF CONSERVATION PRACTICES IN AGRICULTURE IN THE UNITED STATES

INTRODUCTION

Soil and water quality degradation associated with agricultural production have been receiving substantial national attention and are now perceived as environmental problems comparable to other environmental problems such as air quality deterioration and the release of toxic pollutants from industrial sources (Council of Economic Advisors [1997] and National Research Council [1993]). The U.S. Environmental Protection Agency, for example, has identified siltation associated with erosion in rivers and lakes as the second leading cause of water quality impairment, and agricultural production is identified as the leading cause of water quality impairment (Environmental Protection Agency [1995]).

Severe soil degradation from erosion, compaction, or salinization can destroy the productive capacity of the soil.[39] It can also impair water quality from sediment and agricultural chemicals.

Three related causes of water quality impairment are sedimentation, eutrophication, and pesticide contamination. When soil particles and agricultural chemicals wash off a field, they may be carried in runoff until discharged into a water body or stream. Not all agricultural constituents that are transported from a field reach water systems, but a significant portion does, espe-

[39] About 570 million acres (30 percent) of the contiguous United States have a moderate to severe potential for soil and water salinity problems. Saline soils have sufficient soluble salts to

cially dissolved chemicals and the more chemically active, finer soil particles. Once agricultural pollutants enter a water system, they lower water quality and can impose economic losses on water users. These offsite impacts can be substantial. The offsite impacts of erosion are potentially greater than the onsite productivity effects in the aggregate (Foster and Dabney [1995]). Therefore, society may have a larger incentive for reducing erosion than producers have.

If the runoff reaches the water body or stream, soil particles can be suspended in the water, or settle out as sediment, depending on the velocity of the waterflow and the size of the soil particles. In each case, water use can be affected.

Suspended soil particles affect the biologic nature of water systems by reducing the transmission of sunlight, raising surface water temperatures, and affecting the respiration and digestion of aquatic life. The effects on aquatic life, and the reduction in aesthetic quality of recreation sites, can reduce the value of water for recreation uses. Suspended soil particles impose costs on water treatment facilities which must filter out the particles. Suspended soil particles can also damage moving parts in pumps and turbines.

Even when soil particles settle on the bottom of a river or lake, they can cause serious problems for aquatic life by covering food sources, hiding places, and nesting sites. Sedimentation can clog navigation and water conveyance systems like roadside ditches, reduce reservoir capacity, and damage recreation sites. In streambeds, sedimentation can lead to an increase in the frequency and severity of flooding by reducing channel capacity.

The nutrients and pesticides attached to soil particles, or dissolved in runoff, affect water quality in ways that can affect the suitability of water for many uses (Baker [1987]). The most far-reaching impact is eutrophication, abundant growth of algae and rooted vegetation caused by excessive nutrient runoff. As algae dies and decays, it uses oxygen from the surrounding water, lowering the dissolved oxygen levels and altering the size and composition of commercial and recreational sport fisheries. Rooted plants can become a nuisance around marinas and shorelines. Floating algae blooms can restrict light penetration to surface waters and can affect the health, safety, and enjoyment of people using water for recreation. Floating algae can clog intake pipes and filtration systems, increasing the cost of water treatment.

---

adversely affect plant growth. At least 48 million acres of cropland and pastureland are currently affected (Natural Resources Conservation Service [1997]).

Pesticides, which include herbicides, insecticides, and fungicides, create a broad array of impacts. Most notable are effects on aquatic life. Very high concentrations will kill organisms outright. Lower concentrations, more commonly observed, can produce a variety of sublethal effects such as to lower resistance of fish, which makes them susceptible to other stresses (Glotfelty [1987]). Herbicides can hinder photosynthesis in aquatic plants (Schepers [1987]). Pesticides can damage commercial and sport fisheries and make fish dangerous to eat (Herndon [1987]).

To deal with soil degradation issues and water pollution problems associated with conventional agricultural production practices, a number proposals have been advanced.[40] These include conserving and enhancing soil quality[41]

---

[40] Several studies are available that evaluate the improvements in water quality associated with the use of alternative production practices. A few of these are reviewed. Richards and Baker [1998] report on the effort to reduce the eutrophication in Lake Erie that began in the early 1970s. A monitoring station was set up at Bowling Green, Ohio on the Maumee River which feeds into Lake Erie. Agriculture is the dominant land use in the Maumee River basin. Major crops are corn, soybeans, and wheat. Between 1975 and 1995, implementation of no till and reduced tillage increased from less than 5 percent to more than 50 percent of planted acreage. Fertilizer (nitrogen and phosphorous) application rates also changed over the period so it is not possible to quantify precisely the contribution of conservation tillage to the water quality improvements. Nevertheless, water quality changes over the study period were evaluated by conducting trend studies of concentrations and loads. The adoption of conservation tillage in conjunction with the reduced fertilizer application rates led to a reduction in total phosphorous loadings of 24 percent, a reduction in suspended sediments of 19 percent, and a reduction in total Kjeldahl nitrogen of 10 percent. Fawcett et al. [1994] survey the effects of various best management practices on pesticide runoff into surface water and leaching into groundwater. They conclude that no till systems provide a reduction in runoff losses for active pesticide ingredients studied. Average herbicide runoff in no till systems, for example, was 30 percent of the conventional tillage runoff. Additionally, ridge tillage and chisel plow practices are less effective than no till in reducing soil erosion on HEL but are relatively good production practices on less erodible fields. For the various conservation tillage practices relative to conventional tillage, herbicide runoff was 70 less for no till while it was 42 percent less for ridge tillage. With regard to leaching, however, conservation tillage does not fare as well. Increases in infiltration accompanying the use of conservation tillage may result in a greater threat to groundwater from pesticides or nitrate. Preferential flow of water through macropores, which may be more prevalent with no till, can allow water and dissolved solutes or suspended sediment to bypass upper layers of soil. This may transfer pesticides to shallow groundwater or to depths in the soil where biological degradation is slower. It is important, however, to keep this in perspective. Even though there in an increase in the potential leaching risks of certain pesticides associated with conservation tillage, the relative concentrations of pesticides found in surface water are typically greater than concentrations in groundwater.

as the first step towards environmental improvement, increasing nutrient, pesticide, and irrigation use efficiencies in farming systems, increasing the resistance of farming systems to erosion and runoff, and making greater use of field and landscape buffer zones.

Adoption of these proposals depends on the willingness of producers to alter their management and production practices. Producers, however, do not make independent changes in these practices. A change in one component of a management or production practice will generally impact other components of the farming system (Nowak [1992] and Nowak and O'Keefe [1995]).

Inherent links exist between soil quality conservation, improvements in input use efficiency, increases in resistance to erosion and runoff, and the wider use of buffer strips. This links become apparent only if a systems-level approach is taken to analyzing agricultural production systems (National Research Council [1993]). The focus of such an analysis is the farming system, which comprises the pattern and sequence of crops in space and time, the management decisions regarding input and production practices that are used, the management skills, education, and objectives of the producer, the quality of the soil and water, and the nature of the landscape and ecosystems within which agricultural production occurs. An integrated systems approach is necessary for the development of policies and programs to accelerate the adoption of farming systems that are viable for producers, that conserve soil quality, and that do not degrade water quality (Jackson and Piper [1989])

GOVERNMENT POLICIES TO PROMOTE THE ADOPTION OF ALTERNATIVE PRODUCTION PRACTICES

Policies to control soil erosion on agricultural lands and thereby improve water quality have been administered mainly by the U.S. Department of Agriculture. This section describes the policies used by USDA to promote soil conservation.

---

[41] There is some disagreement on precisely how to define soil quality. The reader interested in the debate is referred to, e.g., Soil Quality Institute [1996]. For the purposes here, soil quality is defined as the capacity of a specific kind of soil to function, within natural or managed ecosystem boundaries, to sustain plant and animal productivity, maintain or enhance water and air quality, and support human health and habitation (Karlen et al. [1996]).

The conservation and related water quality programs administered by the U.S. Department of Agriculture (USDA) primarily have been designed to induce the voluntary adoption of conservation practices. The USDA has used a number of policy tools including on-farm technical assistance and extension education, cost-sharing assistance for installing practices preferred, rental and easement payments to take land out of production and place it in conservation uses, R&D for evaluating and improving conservation practices and programs, and compliance provisions that require the implementation of specified conservation practices or the avoidance of certain land use changes if a producer wants to be eligible for Federal agricultural program payments. Regulatory and tax policies have not been part of the traditional voluntary approach of U.S. conservation programs. The USDA policy has been to decrease government involvement in farm operations (Reichelderfer [1990]).

The USDA conservation programs are closely tied to State and local programs. Federal and State agencies cooperate with a system of special-purpose local (county) conservation districts that are authorized by State law to provide education and technical assistance to producers, and county Agricultural Stabilization and Conservation (ASC) committees to handle cost-sharing (Libby [1982]).[42] The system assures that financial support and technical assistance are focused on the set of problems relevant to the geographic region and the national interest. The adoption of an alternative production practice generally does not occur as a consequence of any one specific assistance program (Missouri MSEA [1995]). The USDA has a memorandum of understanding with each conservation district to assist in carrying out a long run program.[43] Conservation districts have proven to be practical organizations through which local producers and the Federal Government can join forces to carry out needed soil conservation practices (Rasmussen [1982]). The demand for information has changed over time. Not long ago, an extension agent was the primary source of information on new technologies (van Es [1984] and

[42] The supervision of these committees was transferred to the Farm Service Agency with the passage of the Federal Agricultural Improvement and Reform Act of 1996. The name of the committees was changed to State Technical Committees.

[43] Many conservation programs to be implemented at the State and local levels require States to submit plans or project proposals and funding needs for Federal approval before actual funds are transferred. For multiyear projects, annual plans of work and documentation of progress are required to receive continued funding. A summary of State programs for erosion control is provided in Magleby et al. [1995].

Hefferman [1984]). Now, however, producers are relying on many additional sources of information, including newspapers, magazines, agrichemical dealers, crop consultants, and the Internet.

The U.S. Department of Agriculture Natural Resources Conservation Service (NRCS), formerly the Soil Conservation Service, provides technical assistance to producers and other land users, including local, State, and Federal agencies that manage publicly owned land. NRCS helps district supervisors and others to draw up and implement conservation plans.

Providing Federal cost-sharing assistance to producers for voluntary installation of approved conservation practices is the responsibility of State and county ASC committees. Through the Agricultural Conservation Program (ACP), funds were allocated among the States through the State ASC committees on the basis of soil and water conservation needs. ACP practices eligible for cost sharing were established by a national review group representing all USDA agencies with conservation program responsibilities, the Environmental Protection Agency, and the Office of Management and Budget. The practices were designed to help prevent soil erosion and water pollution from animal wastes or other nonpoint sources, to protect the productive capacity of farmland, to conserve water, to preserve and develop wildlife habitat, and to conserve energy (Holmes [1987]).[44]

The Secretary of Agriculture can also target critical resource problem areas for financial and technical assistance based on the severity of the problem and the likelihood of achieving improvement. Highly erodible and/or environmentally sensitive cropland has recently been targeted because the greatest net social benefits were expected to be associated with a reduction in soil erosion on these land classifications. Targeting, however, will not guarantee that the net benefits (public and private) of any conservation practice will be positive, because net benefits are a function of site-specific factors.

The policy to take land out of production and place it into conservation uses was first used in the Soil Bank Program of the 1950s, and has been significantly increased in the current Conservation Reserve Program (CRP). The Conservation Reserve Program provides for the USDA to enter into 10-15 year agreements with owners and operators to remove highly erodible and other environmentally sensitive cropland from production. Along with conser-

[44] Authority for ACP was terminated on April 4, 1996, when its functions were subsumed by the Environmental Quality Incentives Program (EQIP).

vation, the CRP originally had a second objective of reducing surplus crop production (Osborn [1996]). The more recent emphasis on CRP, however, has been to provide environmental benefits rather than to control the supply of commodities.

Most of the highly erodible land (HEL) contracted into the CRP had suffered much erosion, organic matter loss, and structural deterioration while it was in cultivated crop production.[45] When lands are returned to grass, their structure and organic matter improve and tend to approach the structure and organic matter content of the original grassland soils (Gebhart et al. [1994]). The degree of soil improvement from 10 years of grass is a function of site-specific factors. As a general rule, the greater the amount of soil structure deterioration from past cultural practices, the more likely that grass management will improve the soil's characteristics. Rasiah and Kay [1994] found that if soils had higher levels of organic matter and other stabilizing materials at the time of grass introduction, the time required for soil structure regeneration was reduced. Soils in the CRP typically fit into the category of degraded soils whose organic matter is lower than that of surrounding soils, because they were primarily allowed into the program based on their highly erodible classification (Lindstrom et al. [1992] and Barker et al. [1996]).

The focus of current conservation research has been on the development of environmentally acceptable and sustainable production practices. Goals of this research are to gain a better understanding of how different soils respond to tillage, what amount of tillage is necessary for optimum crop growth, and what combination of mechanical, chemical, and biological practices are needed to create environmentally sustainable crop production. Many conservation practices have been evaluated by the land grant university experiment stations (Moldenhauer et al. [1994]). USDA has funded several major surveys to provide data to assess the extent and determinants of adoption for particular production practices across a wide range of crops and regions. A more in-depth discussion of conservation-related research and development can be found in Karlen [1990].

The major shift in U.S. conservation policy came in the Food Security Act of 1985 in the form of conservation compliance (Heimlich [1991]). This pro-

[45] Highly erodible land is cropland that has an erodibility index greater than or equal to eight. The technical definition of the erodibility index as well as other relevant terms is presented in the Appendix.

vision provided producers with a basic economic incentive to adopt conservation tillage practices or another acceptable plan because agricultural program payments were linked to the adoption of an acceptable conservation system on highly erodible land. While meeting the conservation provisions remains voluntary, a producer who wants to receive certain agricultural program payments and whose cropland is designated as HEL has no choice but to implement an acceptable conservation plan (Crosswhite and Sandretto [1991]). The conservation compliance provision was innovative because it linked farm program payments (private benefits) to conservation performance (social benefits). In 1982, cultivated HEL accounted for almost 60 percent of the total erosion on U.S. cropland in terms of tons per acre per year while it accounted for only 40 percent of total planted acreage (Magleby et al. [1995]). Requirements for conservation compliance were applied to HEL previously cultivated in any year between 1981 and 1985. Conservation compliance required producers producing crops on HEL to implement and maintain an approved soil conservation system by 1995.[46]

Acceptable conservation plans specify economically viable conservation systems designed to reduce soil erosion. Conservation systems are composed of one or more conservation practices. The 1995 status review of approved conservation systems by the Natural Resources Conservation Service provides the first assessment of fully implemented conservation systems under conservation compliance (Natural Resources Conservation Service [1996]). Although the 1995 status review found over 4,000 different conservation systems (combinations or practices) applied throughout the United States, four conservation systems involving conservation cropping sequences, crop residue use, or a combination of these practices with conservation tillage accounted for approximately half of planted HEL acreage.

The most recent manifestation of agricultural program policy is the Federal Agriculture Improvement and Reform (FAIR) Act of 1996. It modifies

[46] A farmer whose cropland is highly erodible and adopts conservation tillage as part of a conservation plan will benefit by controlling the rate of soil erosion, thereby maintaining the long run productivity of the soil. The significance of this benefit depends on a number of elements including current topsoil depth, erosion rates, and the rate at which the farmer discounts benefits in future years. Conservation tillage has the added benefit, unlike many other conservation practices, in that it can lead to an increase in soil organic matter, an increase in soil moisture, reduced soil compaction, etc. All of these have the potential of enhancing soil quality leading to relatively greater yields in the long run and an increase in the value of a farmer's primary asset, the land.

the conservation compliance provisions of the Food Security Act of 1985 to provide producers with greater flexibility in developing and implementing conservation plans, in self-certifying compliance, and in obtaining variances for problems affecting application of conservation plans. Producers who violate conservation plans, or fail to use a conservation system, on highly erodible land risk loss of eligibility for many payments including production flexibility contract payments. An important aspect of this Act is that in self-certifying compliance, there is no requirement that a status review be conducted for producers who self-certify (Nelson and Schertz [1996]). The FAIR Act also does not differentiate between previously cultivated and uncultivated land thereby eliminating the sodbuster program.[47] Newly cropped highly erodible land may use conservation systems other than the systems previously required under the sodbuster program.

Additionally, the FAIR Act established a new program, the Environmental Quality Incentives Program (EQIP), that incorporated the functions of ACP and some other environmental programs, designed to encourage producers to adopt production practices that reduce environmental and resource problems. Acceptable plans must improve soil, water, and related natural resources including grazing lands, wetlands, and wildlife habitat. EQIP must be carried out to maximize environmental benefits provided per dollar expended. During 1996-2002, the Secretary of Agriculture will provide technical assistance, education, and cost-sharing to producers who enter into 5- to 10-year contracts specifying EQIP conservation programs.

USE OF CONSERVATION PRACTICES AIMED AT SOIL CONSERVATION AND WATER QUALITY IMPROVEMENTS

The adoption of any agricultural technology is a function of many things. The choice of practice is influenced by farm and producer characteristics, attributes of the technology, economic conditions, and policies. For the adoption of conservation practices, some factors are easier to influence than others. For example, ownership characteristics can also influence the adoption of conservation practices. Owner-operators are more likely to have greater flexibility to adopt conservation practices than nonowner-operators who must often get ap-

[47] Note that the sodbuster program was applicable to HEL uncultivated between 1981 and 1985.

proval from the owner before making production practice changes. Conservation practices may tax the managerial skill of the operator (Nowak [1991]). Producers typically make production decisions within a short time frame, a fact that may discourage investment in measures that increase returns only over the long run as may be the case with conservation practices (Office of Technology Assessment [1990] and Tweeten [1995]). Risk-averse producers do not look favorably upon practices that are perceived as being too risky and, in many situations, conservation tillage, for example, is more risky relative to conventional tillage because of the timing and managerial aspects and greater variability in yields. Also, access to capital may depend on risk. There is a relatively greater chance that something might go wrong with conservation tillage and net returns decline (Fox et al. [1991]).

Most conservation policies attempt to influence the use of conservation practices through demonstrating or ensuring that net benefits of adoption are positive. The policies that have had the greatest impact to date have been those focusing on education and technical assistance. The management complexities associated with conservation practices relative to conventional farming are considerable. Farming using conservation practices requires a different approach to soil preparation, fertilizer application, and weed and insect control. Moreover, conservation practices must be designed according to the unique conditions of the region and the specific needs of the individual producer. Consequently, it is not feasible to design conservation practices that can be applied at all locations across the United States or even within a single region. A successful farming system using conservation practices must be developed from a whole-system point of view. Simply stopping tillage, for example, with no other changes in the cropping system increases the potential for problems and failure (Schumacher et al. [1995]).

Reasons posited for the inability of producers to adopt conservation practices even under favorable economic conditions where the net private benefits of adoption would be positive include a lack of information, a high opportunity cost associated with obtaining information, complexity of the production system, a short planning horizon, inadequate management skills, and a limited, inaccessible, or unavailable support system (Nowak [1992]). Westra and Olson [1997] conducted a survey of producers in Scott and Le Sueur Counties in Minnesota. They found that several noneconomic factors impact the decision to use conservation practices, including whether the producer perceived he or she has the requisite management skill and whether a specific conservation practice or set of practices fits in with the producer's production goals and physical setting of the farm. One surprising result from the survey was that 47 percent of the respondents indicated that they knew nothing about conservation practices. The surveyed counties, however, have little acreage designated as HEL.

Education and technical assistance programs funded by the government can be effective in increasing the use of conservation practices (Logan [1990]). Dickey and Shelton [1987], for example, discuss the education programs in eastern Nebraska that targeted specific areas that were susceptible to relatively high soil erosion rates. The program succeeded in increasing the use of conservation tillage by 20 percent and in reducing soil erosion by 20 percent in the target areas.

Education and technical assistance to mitigate the impediments to use of conservation practices associated with real or perceived management inadequacies can come in a variety of forms, not just through extension education and county agents who work for the government. A large amount of information is available through land grant universities, agrichemical dealers, and independent crop consultants. For example, the Washington State University Agriculture Extension Service has prepared the Pacific Northwest Conservation Handbook that addresses virtually all management issues associated with the adoption of conservation practices in the Pacific Northwest (WSU Agriculture Extension Service [1997]). The Agricultural and Biosystems Engineering Department at Iowa State University has produced the publication Conservation Tillage Systems and Management, which addresses management problems associated with conservation tillage in the Midwest (Midwest Plan Service [1992]). Private groups also provide management information. The Conservation Technology Information Center prepares publications such as Conservation Tillage: A Checklist for U.S. Producers (Conservation Technology Information Center [1997]), which address the weed management, insect management, disease management, and nutrient management issues associated with conservation tillage. Thus, there are quite a few publications available in hard copy or through the Internet that address management issues associated with conservation practices.

Agricultural input supply dealers and crop consultants are also good sources of information on the management complexities associated with conservation practices and how to deal with them. Fertilizer and pesticide dealers and crop consultants have consistent access to producers and consequently have the potential for exercising great influence on the production system used (Center for Agricultural Business [1995]). Until recently, there was little information on the influence of dealers and consultants in tillage decisions. Recent survey instruments include questions about whether decisions are influenced by these sources, but there still are no objective studies that quantify the impact of dealers and crop consultants in assisting producers on the use of conservation practices with various other management problems (Wolf [1995]). Education and technical assistance efforts will succeed in inducing adoption of conservation practices only by those producers who can be shown that they will reap net benefits in the long run. For producers with production or resource characteristics for which conservation practices are not profitable, education and technical assistance will not be a sufficient inducement to adopt.

THE USE OF CONSERVATION PRACTICES BY PRODUCERS IN THE U.S.

The foregoing discussion has provided a discussion of the soil and water quality problems associated with conventional agricultural production practices, the various government programs designed to promote the use of conservation practices, and factors which affect the use of conservation practices. This penultimate section will look at the actual use of selected conservation practices designed to mitigate soil erosion and enhance water quality by producers in the United States

(A) DATA

In 1996, the Agricultural Resources Management Study (ARMS) survey was conducted by the U.S. Department of Agriculture. Annual data were collected on fertilizer and pesticide use, tillage systems employed, cropping sequences, whether the cropland is designated as highly erodible, and information on the use of other inputs and production practices. The complete survey covered corn, soybeans, and winter wheat.[48] Only selected States were surveyed, but about 80 percent of the total planted acreage for the respective

[48] Other commodities were also surveyed including cotton, spring wheat, durum wheat, and potatoes. The sample size for these commodities, however, was too small to draw meaningful inferences.

crops was covered. Five tillage systems, including conventional tillage with moldboard plow, conventional tillage without moldboard plow, mulch tillage, ridge tillage, and no till, are defined based on the use of specific tillage implements and their residue incorporation rates (Bull [1993]). (See Chapter 3 for the definition of the various tillage types.) The specific nontillage-related conservation practice questions involved the use of grassed waterways, terraces, contour farming, stripcropping, and the presence of underground outlets and other drainage channels. This is the first time these sorts of non-tillage conservation practice questions have been asked on any U.S. Department of Agriculture-sponsored survey. Consequently, there is no way to identify a trend in the use of these practices. Nevertheless, the survey results provide a good perspective on the extent of use of grassed waterways, terraces, contour farming, stripcropping, and how pervasive is the presence of underground outlets and other drainage channels. The survey results are presented in Table 8.1 for corn, Table 8.2 for soybeans, and Table 8.3 for winter wheat. Additionally, since government soil conservation programs historically have been targeted at land most susceptible to erosion, the survey results are reported separately for highly erodible land and nonhighly erodible land. This will offer some insights into whether the soil conservation message is reaching its intended audience. That is, it will help quantify an answer to the question of whether conservation practices are used relatively more extensively on land that is most in need of their use and, if so, by how much.

(B) Survey Results

The first thing to note is that 33 (9)[49] percent of corn producers on highly erodible land and 59 (11) percent of corn producers on nonhighly erodible land used none of the identified conservation practices including conservation tillage. For soybeans, the percentages are 20 (7) and 41 (9), respectively, for highly erodible and nonhighly erodible land. For winter wheat, the percentages are 61 (8) and 79 (9), respectively, for highly erodible and nonhighly erodible land.

[49] The standard errors of the estimates are in parentheses.

Table 8.1. Cropping Practices Used in Corn Production in the United States for 1996 by Land Classification

Variable	HEL[1] (n=813)[3]	NONHEL[2] (n=2590)
Conventional Tillage (yes=1, no=0)	0.36 (0.04)[4]	0.62 (0.05)
No Till (yes=1, no=0)	0.37 (0.04)	0.14 (0.03)
Mulch Till (yes=1, no=0)	0.26 (0.04)	0.19 (0.02)
Ridge Till (yes=1, no=0)	0.01 (0.01)	0.03 (0.01)
Years of No Till	1.45 (0.11)	0.56 (0.07)
Grassed Waterways (yes=1, no=0)	0.58 (0.04)	0.26 (0.03)
Terraces (yes=1, no=0)	0.34 (0.04)	0.05 (0.02)
Contour Farming (yes=1, no=0)	0.43 (0.03)	0.06 (0.02)
Strip Cropping (yes=1, no=0)	0.07 (0.01)	0.01 (0.01)
Underground Outlets (yes=1, no=0)	0.38 (0.04)	0.49 (0.05)
Other Drainage Channels (yes=1, no=0)	0.12 (0.11)	0.14 (0.02)
Wetland (yes=1, no=0)	0.02 (0.01)	0.04 (0.01)

Table 8.1. Cropping Practices Used in Corn Production in the United States for 1996 by Land Classification (continued)

Residue Cover (percent)	39.24 (1.81)	26.71 (1.41)
Residue Removed (yes=1, no=0)	0.01 (0.01)	0.02 (0.01)
Number of Pesticide Testaments	1.81 (0.03)	1.81 (0.04)
Number of Tillage Operations	4.34 (0.51)	5.54 (0.47)
Number of Trips Across a Field	2.40 (0.16)	2.95 (0.22)
Total Nitrogen Applied (pounds)	134.74 (11.94)	132.14 (11.30)
Total Potassium Applied (pounds)	43.91 (11.11)	51.35 (10.83)
Total Potash Applied (pounds)	47.01 (11.72)	61.44 (11.18)

1 Denotes highly erodible land.
2 Denotes non-highly erodible land.
3 Sample size.
4 Standard errors of the means are in parentheses.
Note that in 1996 there were 79,507,000 acres of corn planted and 73,147,000 acres harvested.
Source: Agricultural Resource Management Study Survey for 1996

Table 8.2. Cropping Practices Used in Soybean Production in the United States for 1996 by Land Classification

Variable	HEL[1] (n=557)[3]	NONHEL[2] (n=2131)
Conventional Tillage (yes=1, no=0)	0.23 (0.03)[4]	0.43 (0.04)
No Till (yes=1, no=0)	0.52 (0.05)	0.28 (0.04)
Mulch Till (yes=1, no=0)	0.23 (0.06)	0.26 (0.06)
Ridge Till (yes=1, no=0)	0.01 (0.01)	0.01 (0.01)
Years of No Till	2.74 (0.44)	0.81 (0.14)
Grassed Waterways (yes=1, no=0)	0.59 (0.06)	0.23 (0.05)
Terraces (yes=1, no=0)	0.30 (0.04)	0.05 (0.01)
Contour Farming (yes=1, no=0)	0.38 (0.04)	0.05 (0.02)
Strip Cropping (yes=1, no=0)	0.01 (0.01)	0.01 (0.01)
Underground Outlets (yes=1, no=0)	0.41 (0.11)	0.53 (0.13)

Table 8.2. Cropping Practices Used in Soybean Production in the United States for 1996 by Land Classification (continued)

Other Drainage Channels (yes=1, no=0)	0.09 (0.01)	0.16 (0.03)
Wetland (yes=1, no=0)	0.02 (0.01)	0.02 (0.01)
Residue Cover (percent)	52.29 (10.00)	36.96 (11.55)
Residue Removed (yes=1, no=0)	0.01 (0.01)	0.03 (0.01)
Number of Pesticide Testaments	1.75 (0.12)	1.80 (0.11)
Number of Tillage Operations	2.06 (0.25)	3.05 (0.33)
Number of Trips Across a Field	1.94 (0.35)	2.83 (0.33)
Total Nitrogen Applied (pounds)	4.19 (0.62)	3.58 (0.34)
Total Potassium Applied (pounds)	18.16 (1.17)	10.71 (1.51)
Total Potash Applied (pounds)	25.44 (1.83)	21.93 (1.96)

1 Denotes highly erodible land. 2 Denotes non-highly erodible land.
3 Sample size; 4 Standard errors of the means are in parentheses.
Note that in 1996 there were 64,205,000 acres of soybeans planted and 63,409,000 acres harvested.
Source: Agricultural Resource Management Study Survey for 1996

From the survey results presented in Tables 8.1, 8.2, and 8.3, conservation tillage is used more extensively on highly erodible land (HEL) than on non-highly erodible land (NONHEL).[50] No till is most widely used in soybeans and corn production on HEL. (Note there is no statistically significant difference between the proportion of soybean producers and corn producers who use no till.) Conventional tillage is still the predominate system used in the production of corn and winter wheat. For corn and soybean producers on HEL who use no till, the production practice has been used a relatively short time period (less than three years). Moreover, soybean producers on HEL have used no till slightly longer than corn producers.

More than half of corn and soybean producers on HEL have grassed waterways in place to reduce soil erosion. This is in contrast to such producers on NONHEL where only about a quarter have grassed waterways. (Note that there is no way to determine from the data the impact of government programs such as EQIP on the installation of this and the other conservation practices.) Moreover, besides conservation tillage, grassed waterways are the dominant type of conservation practice used to control soil erosion and hence improve water quality.

---

[50] Note that all such statements here and elsewhere are based on there being a statistically significant difference between the means. Tests of statistical significance are conducted at the 95-percent level. In drawing conclusions about statistical significance of the difference in the means of two samples, information on the sample size is needed. In the case of the *Agricultural Resources Management Study Survey*, the sample sizes are large enough so that the asymptotic limits of the relevant distributions can be used. The sample sizes, however, are provided in the tables for the interested reader.

Table 8.3. Cropping Practices Used in Winter Wheat Production in the United States for 1996 by Land Classification

Variable	HEL[1] (n=294)[3]	NONHEL[2] (n=604)
Conventional Tillage (yes=1, no=0)	0.64 (0.04)[4]	0.81 (0.05)
No Till (yes=1, no=0)	0.05 (0.01)	0.02 (0.01)
Mulch Till (yes=1, no=0)	0.29 (0.04)	0.16 (0.03)
Ridge Till (yes=1, no=0)	0.01 (0.05)	0.01 (0.04)
Years of No Till	0.09 (0.03)	0.04 (0.01)
Grassed Waterways (yes=1, no=0)	0.16 (0.02)	0.20 (0.03)
Terraces (yes=1, no=0)	0.23 (0.05)	0.30 (0.09)
Contour Farming (yes=1, no=0)	0.27 (0.09)	0.21 (0.07)
Strip Cropping (yes=1, no=0)	0.19 (0.02)	0.05 (0.01)
Underground Outlets (yes=1, no=0)	0.02 (0.01)	0.03 (0.01)

Table 8.3. Cropping Practices Used in Winter Wheat Production in the United States for 1996 by Land Classification (continued)

Other Drainage Channels (yes=1, no=0)	0.05 (0.01)	0.02 (0.01)
Wetland (yes=1, no=0)	0.09 (0.01)	0.08 (0.01)
Residue Cover (percent)	21.91 (1.95)	18.45 (1.56)
Residue Removed (yes=1, no=0)	0.01 (0.01)	0.05 (0.01)
Number of Pesticide Testaments	0.97 (0.14)	0.69 (0.07)
Number of Tillage Operations	5.33 (1.11)	5.51 (1.07)
Number of Trips Across a Field	4.89 (1.48)	4.93 (1.53)
Total Nitrogen Applied (pounds)	48.44 (12.49)	54.36 (11.44)
Total Potassium Applied (pounds)	13.32 (1.00)	16.29 (1.78)
Total Potash Applied (pounds)	1.34 (0.36)	3.37 (0.56)

1 Denotes highly erodible land. 2 Denotes non-highly erodible land.

3 Sample size.

4 Standard errors of the means are in parentheses.

Note that in 1996 there were 51,958,000 acres of winter wheat planted and 39,679,000 acres harvested.

Source: Agricultural Resource Management Study Survey for 1996

About 40 percent of corn and soybean producers on HEL use underground outlets (e.g., drainage tiles). These are primarily used to control sheet and rill erosion and nitrate and pesticides runoff and leaching (David et al. [1997]). Additionally, the use of underground outlets in conjunction with the presence of wetlands on cropland is virtually nonexistent. There is simply not much wetland acreage planted to corn and soybeans. Finally, there is minimal use of underground outlets and other drainage channels used in the production of winter wheat on both HEL and NONHEL.

The use of crop residue to reduce soil erosion is more prevalent on corn and soybean HEL than on NONHEL. That is, the percent residue cover is greater for corn and soybean HEL acreage than for NONHEL acreage. This is not the case with winter wheat, however, where the residue cover is the same for the two land classifications. This is not surprising given the relatively minimal use of conservation tillage in the production of winter wheat.

The number of pesticide treatments is virtually identical for each of the crops for each type of land. Thus, corn, soybean, and winter wheat producers on HEL do not adjust the number pesticide treatments to control for potential leaching and runoff relative to producers on NONHEL.

Nutrient use on HEL versus NONHEL likewise is nearly the same. Thus, for example, total applied nitrogen is not reduced on HEL to mitigate the potential for nitrate leaching and runoff. This, it would appear, is controlled for by underground outlets and other drainage channels, to the extent it is controlled for at all (Kladivko et al. [1991]).

Table 8.4. Combination of Practices Used in the Production of Corn on Highly Erodible Land in 1996

Production Practice	Conservation Tillage (n=470)[1]	Grassed Waterways (n=498)	Terraces (n=242)	Contour Farming (n=376)	Strip Cropping (n=79)	Underground Outlets (n=323)	Other Drainage Channels (n=83)
Conservation Tillage[2]	1.0	0.68 (0.26)[3]	0.70 (0.36)	0.68 (0.29)	0.44 (0.19)	0.61 (0.24)	0.77 (0.34)
Grassed Waterways	0.61 (0.23)	1.0	0.68 (0.31)	0.77 (0.28)	0.62 (0.25)	0.81 (0.31)	0.72 (0.29)
Terraces	0.37 (0.12)	0.49 (0.12)	1.0	0.65 (0.22)	0.35 (0.13)	0.46 (0.12)	0.41 (0.15)
Contour Farming	0.45 (0.12)	0.57 (0.13)	0.83 (0.31)	1.0	0.67 (0.21)	0.59 (0.22)	0.46 (0.18)
Strip Cropping	0.34 (0.11)	0.51 (0.24)	0.21 (0.04)	0.41 (0.17)	1.0	0.54 (0.22)	0.03 (0.01)
Underground Outlets	0.35 (0.10)	0.53 (0.19)	0.51 (0.20)	0.51 (0.22)	0.51 (0.22)	1.0	0.65 (0.24)
Other Drainage Channels	0.54 (0.23)	0.35 (0.13)	0.34 (0.16)	0.23 (0.09)	0.06 (0.02)	0.41 (0.20)	1.0

1 Sample size.
2 Including no till, mulch till and ridge till.
3 Standard errors of the means are in parentheses.

Next, it was previously noted that a systems-level approach is needed for the efficient use of conservation practices. Typically, a single practice is insufficient to effectively control soil erosion and improve water quality. Table 8.4 through Table 8.9 look at the paired use of conservation practices in the production of corn, soybeans, and winter wheat on HEL and NONHEL. An exhaustive discussion of the results is not needed since the tables are fairly easy to interpret and draw inferences from. Just a few of the highlights will be noted. Grassed waterways are used in conjunction with conservation tillage (including mulch tillage, ridge tillage, and no till) on about 60 percent of the HEL acres used in the production of corn and soybeans. This is in contrast to 25 percent of the NONHEL acres for the two crops. Similarly, paired use of other conservation practices is much more prevalent on HEL than on NON-HEL. The use of contour farming in conjunction with stripcropping in the production of soybeans, for example, is used on 44 percent of HEL but on only 8 percent of NONHEL and the use of terraces in conjunction with strip-cropping in the production of corn is used on 21 percent of HEL but on only 5 percent of NONHEL.

The paired use of various conservation practices is less in the production of winter wheat than in the production of corn and soybeans for both land classifications (HEL and NONHEL). There are a number of reasons for this. First, there are relatively more managerial problems associated with the use of conservation tillage in the production of winter wheat than in the production of corn and soybeans (WSU Agriculture Extension Service [1997]). Second, the production system is more complex (National Research Council [1989]). Finally, the support system for winter wheat producers using conservation practices is not as well developed as it is for corn and soybean producers (Zero Tillage Producers Association [1997]).

CONCLUSION

Soil degradation from erosion and the attendant deterioration in water quality associated with agricultural production are significant problems. The use of appropriate conservation practices can mitigate these problems. The government has been promoting, through voluntary means, the use of such practices. There are impediments to their adoption, however, including a lack of information, a high opportunity cost associated with obtaining information, complexity of the production system, a short planning horizon, inadequate management skills, and a limited, inaccessible, or unavailable support system.

Survey data for 1996 for corn, soybean, and winter wheat production show that conservation practices are still not widely adopted, although they are more prevalent on HEL used in the production of corn and soybeans than on NONHEL. Winter wheat remains an anomaly.

There are a number of things that can be done to promote further the increased use of conservation practices. Clearly, there is a need to do so. Among the public policies that can be used to affect the choices of conservation practices are education and technical assistance, financial assistance, research and development, land retirement, and regulation and taxes. These policy options have been addressed previously in Chapter 4 and Chapter 7.

Table 8.5. Combination of Practices Used in the Production of Corn on Non-Highly Erodible Land in 1996

Production Practice	Conservation Tillage (n=470)[1]	Grassed Waterways (n=498)	Terraces (n=242)	Contour Farming (n=376)	Strip Cropping (n=79)	Underground Outlets (n=323)	Other Drainage Channels (n=83)
Conservation Tillage[2]	1.0	0.35 (0.11)[3]	0.31 (0.12)	0.38 (0.15)	0.31 (0.09)	0.33 (0.14)	0.29 (0.12)
Grassed Waterways	0.25 (0.08)	1.0	0.45 (0.13)	0.45 (0.13)	0.39 (0.11)	0.34 (0.09)	0.26 (0.10)
Terraces	0.26 (0.12)	0.23 (0.08)	1.0	0.37 (0.13)	0.07 (0.12)	0.06 (0.02)	0.05 (0.02)
Contour Farming	0.28 (0.11)	0.25 (0.08)	0.43 (0.17)	1.0	0.42 (0.14)	0.05 (0.02)	0.07 (0.02)
Strip Cropping	0.21 (0.09)	0.22 (0.10)	0.05 (0.02)	0.30 (0.02	1.0	0.06 (0.02)	0.01 (0.01)
Underground Outlets	0.44 (0.16)	0.43 (0.18)	0.09 (0.03)	0.13 (0.03)	0.10 (0.04)	1.0	0.22 (0.10)
Other Drainage Channels	0.21 (0.08)	0.14 (0.03)	0.16 (0.05)	0.08 (0.02)	0.10 (0.03)	0.16 (0.05)	1.0

1 Sample size.
2 Including no till, mulch till and ridge till.
3 Standard errors of the means are in parentheses.

Table 8.6. Combination of Practices Used in the Production of Soybeans on Highly Erodible Land in 1996

Production Practice	Conservation Tillage (n=378)[1]	Grassed Waterways (n=351)	Terraces (n=191)	Contour Farming (n=254)	Strip Cropping (n=18)	Underground Outlets (n=234)	Other Drainage Channels (n=47)
Conservation Tillage[2]	1.0	0.75 (0.25)[3]	0.73 (0.31)	0.77 (0.21)	0.55 (0.12)	0.66 (0.30)	0.76 (0.26)
Grassed Water ways	0.58 (0.25)	1.0	0.68 (0.33)	0.78 (0.25)	0.71 (0.26)	0.61 (0.21)	0.69 (0.16)
Terraces	0.49 (0.20)	0.34 (0.12)	1.0	0.63 (0.30)	0.41 (0.12)	0.43 (0.12)	0.48 (0.17)
Contour Farming	0.48 (0.15)	0.50 (0.11)	0.80 (0.21)	1.0	0.52 (0.22)	0.55 (0.16)	0.45 (0.19)
Strip Cropping	0.35 (0.13)	0.54 (0.18)	0.27 (0.08)	0.44 (0.10)	1.0	0.43 (0.14)	0.07 (0.03)
Underground Outlets	0.45 (0.12)	0.42 (0.16)	0.58 (0.17)	0.58 (0.20)	0.47 (0.18)	1.0	0.42 (0.26)
Other Drainage Channels	0.54 (0.13)	0.47 (0.15)	0.41 (0.17)	0.40 (0.11)	0.05 (0.02)	0.39 (0.14)	1.0

1 Sample size.
2 Including no till, mulch till and ridge till.
3 Standard errors of the means are in parentheses.

Table 8.7. Combination of Practices Used in the Production of Soybeans on Non-Highly Erodible Land in 1996

Production Practice	Conservation Tillage (n=960)[1]	Grassed Waterways (n=497)	Terraces (n=118)	Contour Farming (n=161)	Strip Cropping (n=21)	Underground Outlets (n=953)	Other Drainage Channels (n=289)
Conservation Tillage[2]	1.0	0.27 (0.12)[3]	0.39 (0.14)	0.30 (0.13)	0.29 (0.09)	0.63 (0.15)	0.30 (0.12)
Grassed Waterways	0.23 (0.08)	1.0	0.23 (0.07)	0.17 (0.07)	0.16 (0.06)	0.47 (0.19)	0.19 (0.06)
Terraces	0.24 (0.08)	0.16 (0.02)	1.0	0.47 (0.13)	0.15 (0.04)	0.45 (0.02)	0.04 (0.01)
Contour Farming	0.25 (0.01)	0.18 (0.04)	0.49 (0.17)	1.0	0.19 (0.06)	0.25 (0.06)	0.08 (0.02)
Strip Cropping	0.21 (0.07)	0.12 (0.03)	0.11 (0.03)	0.08 (0.01)	1.0 (0.04)	0.19	0.21 (0.04)
Underground Outlets	0.60 (0.15)	0.62 (0.21)	0.54 (0.14)	0.23 (0.11)	0.29 (0.09)	1.0	0.16 (0.03)
Other Drainage Channels	0.25 (0.03)	0.14 (0.03)	0.08 (0.02)	0.12 (0.04)	0.28 (0.05)	0.14 (0.04)	1.0

1 Sample size.
2 Including no till, mulch till and ridge till.
3 Standard errors of the means are in parentheses.

Table 8.8. Combination of Practices Used in the Production of Winter Wheat on Highly Erodible Land in 1996

Production Practice	Conservation Tillage (n=76)[1]	Grassed Waterways (n=62)	Terraces (n=61)	Contour Farming (n=88)	Strip Cropping (n=48)	Underground Outlets (n=17)	Other Drainage Channels (n=17)
Conservation Tillage[2]	1.0	0.29 (0.05)[3]	0.32 (0.16)	0.34 (0.05)	0.10 (0.05)	0.20 (0.05)	0.15 (0.09)
Grassed Waterways	0.09 (0.03)	1.0	0.35 (0.06)	0.31 (0.04)	0.25 (0.06)	0.24 (0.08	0.25 (0.10)
Terraces	0.28 (0.05)	0.40 (0.16)	1.0	0.65 (0.12)	0.05 (0.05)	0.24 (0.10)	0.15 (0.09)
Contour Farming	0.27 (0.05)	0.42 (0.16)	0.76 (0.05)	1.0	0.15 (0.05)	0.08 (0.03)	0.08 (0.06)
Strip Cropping	0.11 (0.03)	0.29 (0.07)	0.04 (0.03)	0.11 (0.03)	1.0	0.17 (0.06)	0.11 (0.04)
Underground Outlets	0.13 (0.03)	0.18 (0.03)	0.22 (0.08)	0.08 (0.03)	0.14 (0.03)	1.0	0.20 (0.07)
Other Drainage Channels	0.12 (0.04)	0.17 (0.04)	0.13 (0.05)	0.08 (0.02)	0.05 (0.03)	0.19 (0.09)	1.0

1 Sample size.

2 Including no till, mulch till and ridge till.

3 Standard errors of the means are in parentheses.

Table 8.9. Combination of Practices Used in the Production of Winter Wheat on Non-Highly Erodible Land in 1996

Production Practice	Conservation Tillage (n=117)[1]	Grassed Waterways (n=90)	Terraces (n=116)	Contour Farming (n=93)	Strip Cropping (n=23)	Underground Outlets (n=19)	Other Drainage Channels (n=34)
Conservation Tillage[2]	1.0	0.12 (0.03)[3]	0.21 (0.03)	0.23 (0.04)	0.11 (0.07)	0.10 (0.07)	0.10 (0.05)
Grassed Waterways	0.13 (0.03)	1.0	0.45 (0.14)	0.41 (0.15)	0.04 (0.05)	0.00	0.16 (0.08)
Terraces	0.24 (0.09)	0.47 (0.14)	1.0	0.63 (0.23)	0.01 (0.02)	0.00	0.03 (0.06)
Contour Farming	0.26 (0.07)	0.44 (0.15)	0.59 (0.04)	1.0	0.02 (0.03)	0.00	0.11 (0.08)
Strip Cropping	0.08 (0.02)	0.02 (0.01)	0.01 (0.01)	0.01 (0.01)	1.0	0.14 (0.11)	0.15 (0.06)
Underground Outlets	0.11 (0.03)	0.00	0.00	0.00	0.12 (0.01)	1.0	0.12 (0.02)
Other Drainage Channels	0.11 (0.02)	0.14 (0.04)	0.01 (0.01)	0.09 (0.02)	0.17 (0.05)	0.15 (0.08)	1.0

1 Sample size.

2 Including no till, mulch till and ridge till.

3 Standard errors of the means are in parentheses.

REFERENCES

Baker, J., "Hydrologic Effects of Conservation Tillage and Their Importance to Water Quality," in T. Logan, J. Davidson, J. Baker, and M. Overcash (eds.), *Effects of Conservation Tillage on Groundwater Quality: Nitrates and Pesticides*, Lewis Publishers, Chelsea, MI, 1987, pp. 113-124.

Barker, J., G. Baumgardner, D. Turner, and J. Lee, "Carbon Dynamics of the Conservation and Wetland Reserve Programs," *Journal of Soil and Water Conservation*, Vol. 51 (1996), pp. 340-346.

Bosch, D., Z. Cook, and K. Fuglie, "Voluntary Versus Mandatory Agricultural Policies to Protect Water Quality: Adoption of Nitrogen Testing in Nebraska," *Review of Agricultural Economics*, Vol. 17 (1995), pp. 13-24.

Center for Agricultural Business, *Purdue/Top Commercial Producer Study*, Department of Agricultural Economics, Purdue University, West Lafayette, IN, 1995.

Conservation Technology Information Center, *Conservation Tillage: A Checklist for U.S. Producers*, West Lafayette, IN, 1997.

Council of Economic Advisors, *Economic Report of the President*, U.S. Government Printing Office, Washington, DC, 1997.

David, M., D. Kovacic, R. Cooke, and L. Gentry, *Advancing Sustainable Uses of Resources*, University of Illinois Agricultural Extension Service, Urbana, IL, 1997.

Dickey, E., and D. Shelton, "Targeted Education Programs to Enhance the Adoption of Conservation Practices," *Optimum Erosion Control at Last Cost*, American Society of Agricultural Engineers, St. Joseph, MI, 1987.

Fawcett, R., B. Christensen, and D. Tierney, "The Impact of Conservation Tillage on Pesticide Runoff into Surface Water: A Review and Analysis," Journal of Soil and Water Conservation, Vol. 49 (1994), pp. 126-135.

Foster, G., and S. Dabney, "Agricultural Tillage Systems: Water Erosion and Sedimentation," *Farming for a Better Environment*, Soil and Water Conservation Society, Ankeny, IA, 1995.

Fox, G., A. Weersink, G. Sarwar, S. Duff, and B. Deen, "Comparative Economics of Alternative Agricultural Production Systems: A Review," *Northeastern Journal of Agricultural and Resource Economics*, Vol. 20 (1991), pp. 124-142.

Gebhart, D., H. Johnson, H. Mayeux, and H. Polley, "The CRP Increases Soil Organic Carbon," *Journal of Soil and Water Conservation*, Vol. 49 (1994), pp. 488-492.

Glotfelty, D., "The Effects of Conservation Tillage Practices on Pesticide Volatilization and Degradation," *Effects of Conservation Tillage on Groundwater Quality*, Lewis Publishers, Chelsea, MI, 1987.

Hefferman, W., "Assumptions of the Adoption/Diffusion Model and Soil Conservation," in B. English, J. Maetzold, B. Holding, and E. Heady (eds.) *Agricultural Technology and Resource Conservation*, Iowa State University Press, Ames, IA, 1984.

Heimlich, R., "Soil Erosion and Conservation Policies in the United States," in N. Hanley (ed.) *Farming and the Countryside: An Economic Analysis of External Costs and Benefits*, CAB International Publishers, Miami, FL, 1991.

Herndon, L., "Conservation Systems and Their Role in Sustaining America's Soil, Water, and Related Natural Resources," *Optimum Erosion Control at Least Cost*, American Society of Agricultural Engineers, St. Joseph, MI, 1987.

Holmes, B., Legal Authorities for Federal (USDA), *State, and Local Soil and Water Conservation Activities: Background for the Second RCA Appraisal*, U.S. Department of Agriculture, Washington, DC, 1987.

Jackson, W., and J. Piper, "The Necessary Marriage between Ecology and Agriculture," *Bulletin of the Ecological Society of America*, Vol. 70 (1989), pp. 1591-1593.

Karlen, D., "Conservation Tillage Research Needs," *Journal of Soil and Water Conservation*, Vol. 45 (1990), pp. 365-369

Karlen, D., N. Wollenhaupt, D. Erbach, E. Berry, . Swan, N. Eash, and J. Jordal, "Long Term Tillage Effects on Soil Quality," *Soil and Tillage Research*, Vol. 32 (1996), pp. 313-327.

Kladivko, E., G. Van Scoyoc, E. Monke, K. Oates, W. Pask, "Pesticide and Nutrient Movement into Subsurface Tile Drains on a Silt Loam in Indiana," *Journal of Environmental Quality*, Vol. 20 (1991), pp. 213-227.

Libby, L., "Interaction of RCA with State and Local Conservation Programs," in H. Halcrow, E. Heady, and M. Cotner (eds.), *Soil Conservation Policies Institutions, and Incentives*, Soil Conservation Society of America, Ankeny, IA, 1982.

Lindstrom, M., T. Schumacher, A. Jones, and C. Gantzer, "Productivity Index Model for Selected Soils in North Central United States," *Journal of Soil and Water Conservation*, Vol. 47 (1992), pp. 491-494.

Logan, T., "Agricultural Best Management Practices and Groundwater Protection," *Journal of Soil and Water Conservation*, Vol. 45 (1990), pp. 201-206.

Magleby, R., C. Sandretto, W. Crosswhite, and C.T. Osborn, *Soil Erosion and Conservation in the United States, AIB 718*, Economic Research Service, U.S. Department of Agriculture, Washington, DC, 1995.

Moldenhauer, W., W. Kemper, and G. Langdale, "Long Term Effects of Tillage and Crop Residue Management," in G. Langdale and W. Moldenhauer (eds.), *Crop Residue Management to Reduce Erosion and Improve Soil Quality - Southeast, CR-39*, Agricultural Research Service, U.S. Department of Agriculture, Washington, DC, 1994.

Missouri Management Systems Evaluation Area, *A Farming Systems Water Quality Report, Research and Education Report*, Missouri MSEA, 1995.

National Research Council, *Alternative Agriculture*, National Academy Press, Washington, DC, 1989.

National Research Council, *Soil and Water Quality*, National Academy Press, Washington, DC, 1993.

Natural Resources Conservation Service, *A Geography of Hope*, U.S. Department of Agriculture, Washington, DC, 1997.

Nelson, F., and L. Schertz, *Provisions of the Federal Agriculture Improvement and Reform Act of 1996*, Economic Research Service, U.S. Department of Agriculture, Washington, DC, 1996.

Nielson, J., *Targeting Erosion Control: Delivering Technical and Financial Assistance to Producers*, Agricultural Research Service, U.S. Department of Agriculture, Washington, DC, 1985.

Nowak, P., "Why Producers Adopt Production Technology," in *Crop Residue Management for Conservation*, Proceedings of a National Conference, Soil and Water Conservation Society, Ankeny, IA, 1991.

Nowak, P., "Why Producers Adopt Production Technology," *Journal of Soil and Water Conservation*, Vol. 47 (1992), pp. 14-16.

Nowak, P., and G. O'Keefe, "*Producers and Water Quality: Local Answers to Local Issues*," Draft Report submitted to the U.S. Department of Agriculture, Washington, DC, September 1995.

Osborn, T., "Conservation," *Provisions of the Federal Agricultural Improvement Act of 1996*, Economic Research Service, U.S. Department of Agriculture, Washington, DC, 1996.

Rasiah, R., and B. Kay, "Characterizing the Changes in Aggregate Stability Subsequent to Introduction of Forages," *Soil Science Society of America Journal*, Vol. 58 (1994), pp. 935-942.

Rasmussen, W., "History of Soil Conservation, Institutions and Incentives," in H. Halcrow, E. Heady, and M. Cotner (eds.), *Soil Conservation Policies Institutions, and Incentives*, Soil Conservation Society of America, Ankeny, IA, 1982.

Reichelderfer, K., "Land Stewards or Polluters? The Treatment of Producers in the Evaluation of Environmental and Agricultural Policy," a paper presented at the conference *Is Environmental Quality Good for Business?*, Washington, DC, 1990.

Richards, R., and D. Baker, *Twenty Years of Change: The Lake Erie Agricultural Systems for Environmental Change (LEASEQ) Project*, Water Quality Laboratory, Heidelberg College, Tiffin, OH, 1998.

Saha, A., H.A. Love, and R. Schwart, "Adoption of Emerging Technologies Under Output Uncertainty," *American Journal of Agricultural Economics*, Vol. 76 (1994), pp. 836-846.

Schepers, J., "Effects of Conservation Tillage on Processes Affecting Nitrogen Management," *Effects of Conservation Tillage on Groundwater Quality*, Lewis Publishers, Chelsea, MI, 1987.

Schumacher, T., M. Lindstrom, M. Blecha, and R. Blevins, "Management Options for Lands Concluding their Tenure in the Conservation Reserve Program," in R. Blevins and W. Moldenhauer (eds.) *Crop Residue Management to Reduce Erosion and Improve Soil Quality - Appalachia and Northeast*, Agricultural Research Service, U.S. Department of Agriculture, Washington, 1995.

Soil Quality Institute, *The Soil Quality Concept*, Natural Resources Conservation Service, U.S. Department of Agriculture, Ames, IA, 1996.

Tweeten, L., "The Structure of Agriculture: Implications for Soil and Water Conservation," *Journal of Soil and Water Conservation*, Vol. 50 (1995), pp. 347-351.

U.S. Congress, Office of Technology Assessment, Beneath the Bottom Line: *Agricultural Approaches to Reduce Agrichemical Contamination of*

Groundwater, OTA-F-418, U.S. Government Printing Office, Washington, DC, November 1990.

U.S. Department of Agriculture, Natural Resources Conservation Service, *1995 Final Status Review Results*, Washington, DC, 1996.

U.S. Environmental Protection Agency, Office of Water, *National Water Quality Inventory - 1994 Report to Congress*, Office of Water, Washington, DC, 1995.

van Es, J. "Dilemmas in the Soil and Water Conservation Behavior of Producers," in B. English, J. Maetzold, B. Holding, and E. Heady (eds.) *Agricultural Technology and Resource Conservation*, Iowa State University Press, Ames, 1984.

Washington State University Agriculture Extension Service, *Pacific Northwest Conservation Tillage Handbook*, Washington State University, Pullman, WA, 1997.

Westra, J., and K. Olson, *Producers' Decision Processes and the Adoption of Conservation Tillage*, Department of Applied Economics, University of Minnesota, Minneapolis, MN, 1997.

Wolf, S., "Cropping Systems and Conservation Policy: The Roles of Agrichemical Dealers and Crop Consultants," Journal of Soil and Water Conservation, Vol. 50 (1995), pp. 263-270.

CHAPTER 8 APPENDIX

The Universal Soil Loss Equation (USLE) is

$$A = R * K * f(L,S) * C * P$$

where

A is the computed soil loss per unit area, expressed in the units selected for K and for the periods selected for R. In practice, these are usually so selected that they compute A in tons per acre per year;

R, the rainfall and runoff factor, is the number of the rainfall erodibility index units plus a factor for runoff from snow melt or applied water where such runoff is significant;

K, the soil erodibility factor is the soil loss rate per erodibility index unit for a specified soil as measured on a unit plot, which is defined as a 72.6-ft length of uniform 9-percent slope continuously in clean-tilled fallow;

L, the slope length factor, is the ratio of soil loss from the field slope to that from a 72.6-ft length of uniform 9-percent slope continuously in clean-tilled fallow;

S, the slope steepness factor, is the ratio of soil loss from the field slope gradient to that from a 72.6-ft length of uniform 9-percent slope continuously in clean-tilled fallow;

C, the cover and management factor, is the ratio of soil loss from an area with specified cover and management to that from an identical area in tilled continuous cover; and

P, the support practice factor, is the ratio of soil loss with a supporting practice like contouring, strip cropping, or terracing to that with straight-row farming up and down the slope.

Note that f(L,S) indicates a functional relationship between L and S.

Highly erodible land (HEL) is land determined to have an inherent erosion potential of over 8 times its soil loss tolerance (T) level. Determination is made by calculating the erodibility index (EI) for both water and wind erosion. If the EI for either water or wind is greater than 8, then the soil is classified as HEL.

The erodibility index is a number showing how many times the inherent erosion potential is of the soil loss tolerance (T) level. For water (sheet and rill) erosion, the number is calculated as

$$EI = R * K * L * S / T$$

where

R, the rainfall and runoff factor, is the number of the rainfall erodibility index units plus a factor for runoff from snow melt or applied water where such runoff is significant;

K, the soil erodibility factor is the soil loss rate per erodibility index unit for a specified soil as measured on a unit plot, which is defined as a 72.6-ft length of uniform 9-percent slope continuously in clean-tilled fallow;

L, the slope length factor, is the ratio of soil loss from the field slope to that from a 72.6-ft length of uniform 9-percent slope continuously in clean-tilled fallow; and

S, the slope steepness factor, is the ratio of soil loss from the field slope gradient to that from a 72.6-ft length of uniform 9-percent slope continuously in clean-tilled fallow.

The soil loss tolerance level (T) is the maximum rate of annual soil erosion that may occur and still permit a high level of crop productivity to be obtained economically and indefinitely. Most values for cropland in the United States are between 3 and 5 tons per acre per year.

CHAPTER 9

THE IMPACT OF ENERGY ON THE ADOPTION OF CONSERVATION TILLAGE IN THE UNITED STATES

INTRODUCTION

Soil erosion associated with agricultural production practices can impose significant costs. The most widespread offsite erosion-related problem is impairment of water resource use (National Research Council [1993]). The Environmental Protection Agency has identified siltation associated with erosion in rivers and lakes as the second leading cause of water quality impairment, and agricultural production is identified as the leading cause of water quality impairment (Environmental Protection Agency [1995]).

Conservation tillage practices that leave substantial amounts of crop residue evenly distributed over the soil surface defend against the potential of rainfall's kinetic energy to generate sediment and increase water runoff. Several field studies (e.g., Baker and Johnson [1979], Glenn and Angle [1987], Hall et al. [1984], and Sander et al. [1989]) conducted under natural rainfall on highly erodible land (14 percent slope) have compared erosion rates between tillage systems. Compared with moldboard plowing, conservation tillage generally reduced soil erosion by 50 percent or more. Specifically, Hall et al. [1984] report that when compared with conventional tillage, conservation tillage reduced runoff by 86.3 to 98.7 percent, soil losses by 96.7 to 100 percent, and the herbicide cyanazine losses by 84.9 to 99.4 percent. Glenn and Angle [1987] report a 27 percent less total runoff of water and the herbicides atrazine and simazine with conservation tillage versus conventional tillage systems.

Increased surface residues also filter out and trap sediment and sediment absorbed agricultural chemicals (pesticides and fertilizer) and result in a cleaner runoff (Onstad and Voohees [1987]). The increase in organic matter associated with crop residue management intercepts the chemicals and holds them in place until they are used by the crop or degrade into inert components (Dick and Daniel [1987] and Wagnet [1987]). The presence of increased crop residue typically reduces the volume of contaminants entering surface water by constraining runoff (including dissolved chemicals and sediment) and enhancing filtration (Baker [1987] and Fawcett [1987]).

The adoption of conservation tillage has been increasing in the United States in response to growing concerns about the impact of agricultural production on the environment. There is an identifiable upward trend until the last few years, which show no discernible change. A longer term perspective can be obtained from Figure 9.1. The adoption of conservation tillage increased from 1 percent of planted acreage in 1963 to 37 percent of planted acreage in 1997 (Schertz [1988] and Conservation Technology Information Center [annual]).

Figure 9.1. Percent of Planted Acres on Which Conservation Tillage is Used in the United States

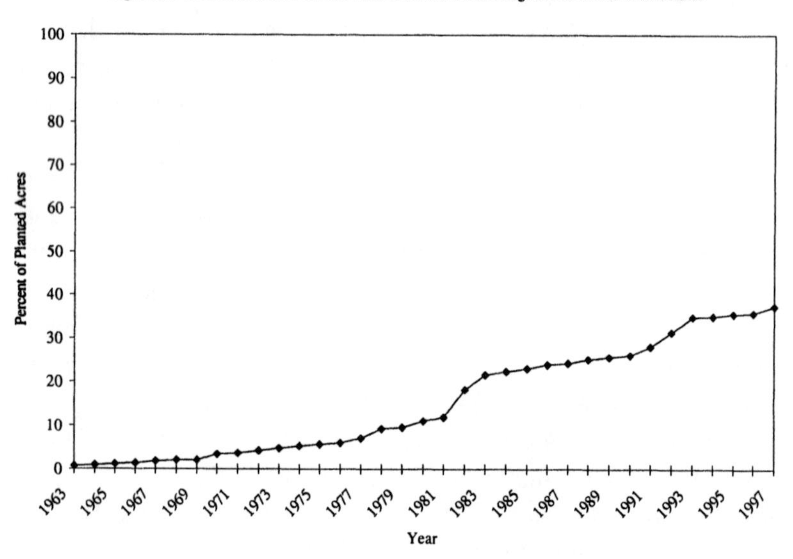

As noted, the adoption of conservation tillage practices frequently reduces the negative impacts associated conventional tillage systems. These negative impacts include energy use, soil erosion, leaching and runoff of agricultural chemicals, and carbon emissions. The relationship between energy and the adoption of conservation tillage is of special importance and is the subject of what follows.[51] Before exploring this, however, some background is needed.

Background

Farmers in general tend to make production practice changes slowly. The adoption process generally can be viewed as having five stages (Nowak [1992] and Nowak and O'Keefe [1992, 1995]). Initially, farmers are unaware of a new practice (stage 1). They become aware of new practices through various sources, including neighbors, farm publications, mass media, extension agents, chemical dealers, and crop consultants (stage 2). Farmers then evaluate the practice in terms of their own operation through educational sources such as demonstration projects, talking with agents, and talking with neighbors who have tried the practice (stage 3). Farmers may then test the practice on part of their farm (stage 4). The ability of a practice to be tested on part of the farm enhances its potential for adoption. Finally, full adoption occurs if the practice is found to be better that what they are currently doing (stage 5).

A variety of economic, demographic, geographic, and policy variables have been identified that affect the adoption and adoption of conservation tillage in the United States.[52] The rate of adoption (diffusion) of a new technol-

[51] Energy for tillage operations accounts for approximately 3 percent of total farm energy use while energy use for fertilizer and pesticides applications accounts for 42 percent and 4 percent, respectively, of total farm energy use in the United States (Torgerson et al. [1987]).

[52] Because of the nature of the data available and their limitations, however, it is generally not possible to quantify precisely the impact that these factors have on conservation tillage use. Among the more serious data limitations are the availability of only a relatively short time series, incomplete measurement of all of the relevant variables for the times series that are available, measurement error in the data, absence of panel data, and difficulty in separating the impact of government policy (conservation compliance) from varying climatic and economic conditions. With regard to the first issue, consistent data on conservation tillage date only to the late 1980s. Six or seven years' worth of annual observation is typically inadequate to estimate a formal structural model when there are more that just a few factors affecting the dependent variable (the relative use of conservation tillage). Of the time series data that are available, information on coincident climatic conditions and relevant economic variables are not collected in any sort of useable way. The changing definition of conservation tillage over the longer period has introduced measurement error of the dependent variable. The absence of panel data makes it

ogy -- e.g., conservation tillage -- determines the rate of technological change. The first empirical assessment of the diffusion of a new technology was applied to hybrid corn (Griliches [1957]). The diffusion follows an innovation cycle. The cycle starts with efficient producers first introducing the new technology that requires a threshold level of technical skill for profitable use. As skill levels of other farmers increase through experience, the new technology is more widely adopted. The time path of adoption of the new technology can be analytically derived as a function of the distribution of technical ability among producers and the rate of change in technical skill (Kislev and Schori-Bachrach [1973]).

Adoption is also a function of exogenous factors, and these factors will retard or accelerate the rate of adoption. Investment costs associated with the adoption of the new technology will have an important influence on a farmer's choice. Government policy in the form of conservation compliance is an example of an exogenous factor that would be expected to accelerate the rate of adoption of conservation tillage (Batte [1993]).[53] Yet another exogenous consideration is what is nominally referred to as learning by doing (Alchien [1959]). When specialized management skills are required for production, owner/operators will gain proficiency with experience; that is, they learn by doing.

In the context of the diffusion of conservation tillage as a new technology (production practice), a sizeable number of studies are available that provide some insights into the important factors that affect the adoption of conservation tillage. Because there is considerable redundancy in the results of the studies, an exhaustive survey will not be provided. Pagoulatos et al. [1989] using an erosion-damage function analysis for corn grown in Kentucky found that the decision to convert to conservation tillage from conventional tillage is dependent on the price of output, the discount rate (with a higher discount rate

---

impossible to monitor the behavior of any group of farmers over time. Consequently, because of these data limitations, in the review that follows, many of these factors affecting conservation tillage use will be discussed, but it is not feasible to quantify their relative importance in influencing the trend in the use of conservation tillage.

[53] The Food Security Act of 1985 targeted highly erodible land under the conservation compliance provision. This provisions stipulated that farmers of such land who did not implement approved conservation plans would lose eligibility for U.S. Department of Agriculture program benefits. Under conservation compliance, farmers with highly erodible cropland had to have a plan in place by January 1, 1990 and that plan implemented by January 1, 1995. Many different conservation tillage practices were acceptable.

leading to a slower adoption of conservation tillage), and the capital cost of conversion. Large capital costs for new machinery serve as a deterrent to adoption of conservation tillage.

Gray et al. [1997] used a simulation model to compare the adoption of conservation tillage systems to conventional tillage systems for wheat production in western Canada. Crop yield and the price of the burndown herbicide (the herbicide used to eliminate vegetation prior to planting) were key determinants to the short run profitability of adopting conservation tillage.

Carter and Kunelius [1990] analyzing data from Atlantic Canada found that some soil types are simply not suitable for conservation tillage.[54] The soils require a high degree of cultivation to maintain their structure and regular tillage to ensure adequate crop productivity. Moreover, climatic constraints such as a short growing season, cold temperatures, and excessive precipitation can influence the choice of a conservation tillage system.

Batte et al. [1993] find that commercial farms in Ohio in 1992 tended to operate with a single system. Thus, farms classified as no till used no till on 85 percent of planted acreage while conventional tillage farms used moldboard plowing on 80 percent of their acreage. Also, farms using conservation tillage tended to be substantially larger than farms using conventional tillage.

The greater risk associated with the adoption of conservation tillage has been shown to be a deterrent to the adoption of conservation tillage in a number of studies. Risk in these studies is typically defined as variability in yields or variability in net returns. Thus, Mikesell et al. [1988] using a simulation model evaluated the expected net returns and risk of alternative tillage systems for a 640-acre grain farm in northeastern Kansas. Conservation tillage systems had slightly higher expected incomes but were more risky. A risk-averse farmer would prefer conventional tillage to conservation tillage. Williams et al. [1988] using a simulation model found that conservation tillage used in grain sorghum production had higher expected net revenues but greater risk than conventional tillage.

Westra and Olson [1997] estimated a structural model based on survey responses for farmers in two counties in Minnesota. Their results suggest that larger farms are more likely to use conservation tillage. Also, if the owner/operator is relatively more concerned about erosion, the probability of adopting conservation tillage is greater. The greater complexity associated

[54] The soil types are not identified.

with the adoption of conservation tillage requiring greater management skill is identified as a deterrent to the adoption of conservation tillage.

Even though it is recognized as being important, one factor typically absent from explicit consideration in empirical examinations of conservation tillage adoption and use is energy. The variability in energy costs associated with conservation tillage is a consequence of the change in the number of trips across a field associated with conservation tillage relative to conventional tillage. Fewer tillage operations result in fewer trips across the field (Frye [1995] and Griffith et al. [1977]).[55] In addition, however, while the number of trips across a field to apply fertilizer does not vary between tillage systems,[56] the number of pesticide treatments and, hence, the number of trips across the field to apply pesticides, is greater for corn, soybeans, and spring wheat grown under conservation tillage than under conventional tillage for some years between 1990 and 1995 (Table 9.1). Associated with a larger number of treatments is an increase in energy use. Thus, while energy use has typically been considered to be a significant factor in the tillage adoption decision,[57] important questions are whether energy considerations in the aggregate de facto affect the adoption of conservation tillage and, if so, to what extent do they impact the adoption of conservation tillage in the United States. Both of these questions will be subsequently addressed.

---

[55] It has been estimated that energy use for tillage operations is reduced by between 3.4 and 9.8 percent when conservation tillage is used instead of conventional tillage (Frye [1984]).

[56] Based on National Agricultural Statistical Service/Economic Research Service (NASS/ERS) *Cropping Practices Survey* data, the average for corn for 1995 was 1.92 (0.03) trips per field for conservation tillage and 1.90 (0.02) trips per field for conventional tillage. (Standard errors of the means are in parentheses.) The average for soybeans, winter wheat, spring wheat, and durum wheat for conservation tillage is 0.31 (0.02), 1.32 (0.05), 1.31 (0.09), and 1.46 (0.11), respectively. The average for soybeans, winter wheat, spring wheat, and durum wheat for conventional tillage is 0.32 (0.02), 1.40 (0.03), 1.29 (0.05), and 1.40 (0.09), respectively. Consequently, there are no statistically significantly differences in energy costs associated with the application of fertilizer between conservation tillage and conventional tillage. The same results also hold for 1990-94.

[57] This is illustrated by a provision in the Energy Security Act of 1980 that amended the Agricultural Conservation Program (ACP) authorizing legislation to include conservation tillage and other ostensibly more energy efficient conservation practices (U.S. Congress [1980]). The ACP provided cost-sharing assistance to farmers for implementing conservation practices designed to restore and improve soil fertility and minimize erosion caused by wind and water.

Energy and Conservation Tillage Use

To address the question of whether energy considerations affect the adoption of conservation tillage in the United States, one approach is to assess the impact of the price of energy on the adoption of conservation tillage. This will be done in what follows.

(A) Methodology

One direct way to determine whether there is a relationship between the real price of energy and the adoption of conservation tillage in the United States is to examine empirically whether there is an identifiable long run equilibrium relationship between the two variables. Although the variables may drift away from equilibrium for a while, economic forces may be expected to act so as to restore equilibrium. This sort of inherent long run relationship leads naturally to the concept of cointegration between the two variables. If two variables are cointegrated, then they must obey an equilibrium relationship in the long run, although they may diverge substantially from equilibrium in the short run. The concept of cointegration is fundamental to the understanding of a long run relationship among economic time series. It will be discussed below.

Table 9.1. Average Number of Pesticide Treatments by Tillage Practice - 1995

Commodity	Conventional Tillage	Conservation Tillage
Corn		
1990	1.63 (0.02)1	1.69 (0.03)
1991	1.69 (0.02)	1.78 (0.04)
1992	1.71 (0.02)	1.82 (0.03)
1993	1.67 (0.02)	1.77 (0.02)
1994	1.72 (0.02)	1.82 (0.02)
1995	1.79 (0.03)	1.93 (0.03)
Soybeans		
1990	1.49 (0.02)	1.51 (0.03)
1991	1.54 (0.03)	1.63 (0.04)
1992	1.57 (0.03)	1.63 (0.03)
1993	1.47 (0.02)	1.62 (0.02)
1994	1.67 (0.02)	1.77 (0.02)
1995	1.64 (0.03)	1.87 (0.04)

Table 9.1. Average Number of Pesticide Treatments by Tillage Practice – 1995, (continued)

Winter Wheat		
1990	0.41 (0.02)	0.55 (0.05)
1991	0.39 (0.02)	0.36 (0.05)
1992	0.39 (0.02)	0.32 (0.04)
1993	0.50 (0.02)	0.49 (0.05)
1994	0.53 (0.02)	0.55 (0.05)
1995	0.76 (0.02)	0.75 (0.05)
Spring Wheat		
1990	1.26 (0.05)	1.42 (0.14)
1991	1.17 (0.04)	1.20 (0.07)
1992	1.14 (0.05)	1.03 (0.07)
1993	1.19 (0.06)	1.12 (0.08)
1994	1.24 (0.07)	1.18 (0.05)
1995	1.07 (0.02)	1.17 (0.03)
Durum Wheat		
1990	1.48 (0.09)	1.52 (0.12)
1991	1.38 (0.09)	1.47 (0.10)
1992	1.45 (0.11)	1.27 (0.09)
1993	1.67 (0.27)	1.56 (0.33)
1994	1.38 (0.07)	1.42 (0.10)
1995	1.41 (0.09)	1.50 (0.14)

1 The values in parentheses are the standard errors.
Source: Cropping Practices Survey - 1990-1995

Cointegration exists between two nonstationary time series (a nonstationary time series is one whose mean and/or variance changes over time and whose covariance between values at two points in time of the same distance vary when alternative time points are considered) that are both integrated of the same order, I(d), if there is a linear combination of the two series which is itself stationary. By definition, a variable is integrated of order d if a dth difference of the series is stationary. A stationary series is denoted as I(0). Two economic series, sx and sy, are cointegrated if the relationship between the two can be written as

(2) $zt = sxt - \kappa\, syt$

where zt is I(0) and κ is the cointegrating constant. The equilibrium error process, zt, represents the deviation of the series for x and y away from the long run equilibrium.[58]

One test for cointegration involves ordinary least squares (OLS) estimation of the following static cointegrating regression:

$$(3) \qquad syt = \upsilon + \phi\ sxt + ut$$

where υ and ϕ are coefficients to be estimated and ut is the stochastic term. The null hypothesis of no cointegration is rejected if ϕ is statistically significant and if ut is I(0). This test is not very powerful, however, in the event that ut is stationary but highly serially correlated (Jenkinson [1986]).

Error correction models provide an alternative test for cointegration. This approach models both the short run dynamics and the long run equilibrium between variables suggested by economic theory. It is also possible to draw causal inferences on the basis of error correction models. For economic series that are cointegrated, causality must run in at least one direction since one economic series can be used to help forecast the other.[59] (Granger [1986, 1988] discusses the causal implications of cointegration.)

According to the Granger Representation Theorem (Granger and Weiss [1986], Engle and Granger [1987], and Engle and Yoo [1987]), if two series are cointegrated, then there is an error correction model (ECM) of the following form:

$$(4)\ (1\text{-}L)\ syt = \text{-}p1\ zt\text{-}1 + A(L)\ (1\text{-}L)\ syt + B(L)\ (1\text{-}L)\ sxt + u1t$$
$$(5)\ (1\text{-}L)\ sxt = \text{-}p2\ zt\text{-}1 + C(L)\ (1\text{-}L)\ sxt + D(L)\ (1\text{-}L)\ syt + u2t$$

where zt-1 = sx(t-1) - κ sy(t-1), p1 and p2 are nonzero parameters, and u1t and u2t are both I(0). The one sided lag polynomials A(L), B(L), C(L), and D(L) are stable such that the roots of the associated polynomial are outside the unit circle. (A one sided lag polynomial is defined whereby the current value of the

[58] For a further discussion of cointegration and of alternative testing procedures, the interested reader is referred to the excellent collection of papers in the *Oxford Bulletin of Economics and Statistics*, Vol. 38, No. 3 (1986).

[59] Since the error correction term is a function of the levels of the economic series, lagged values of the other series are significant in explaining movements in the series that serves as the dependent variable when the lagged error correction term is significant.

variable of interest is solely a function of past values of that variable plus, perhaps, a constant.) The term L is the lag operator where LkWt = Wt-k. If two economic series are cointegrated, the coefficient on the error correction term, zt-1, must be statistically significant in at least one of the error correction equations (i.e., equations (4) and (5)).

Causal inferences are based on the statistical significance of p1 and p2 and the elements in B(L) and D(L). For example, p1 and the elements in B(L) equal to zero supports the conclusion that the real price of crude oil does not Granger-cause fluctuations in the adoption of conservation tillage.[60]

The foregoing describes the methodology that will be used to investigate the cointegration between a change in the price of crude oil and unemployment.

(B) Data

To assess whether there is an identifiable relationship between the adoption of conservation tillage and the price of energy, data covering the period 1963 to 1997 were used. The untransformed crude oil price data, which is used as a proxy for the price of energy[61] and which represent the composite refiner acquisition cost, were taken from Council of Economic Advisors [1979] and the Energy Information Administration [1997]. The price data are deflated by the gross domestic product implicit price deflator. These data were obtained from the Bureau of Economic Analysis of the U.S. Department of Commerce. The conservation tillage use data were obtained from Schertz

[60] Granger [1988] discusses the possibility of viewing the causal impact of the error correction term as occurring at low frequencies (i.e., in the long run). For example, p_1 different from zero would indicate long run causality from crude oil price volatility to changes in the unemployment rate, while B(L) different from zero would indicate short run causality. While such an interpretation is attractive, Granger warns that it is unclear whether such a view is justified until analysis similar to that of Geweke [1984] is completed for the error correction test being considered here. Such an effort would explore the frequency decomposition of the error correction term.

[61] Actually, the concern is with the impact of the prices of refined petroleum products, which are factors of production used in producing various agricultural commodities. But since there is a high degree of collinearity between the price of crude oil and the prices paid by farmers for diesel fuel, gasoline, and liquefied petroleum gas (simple correlations are 0.93, 0.97, and 0.96 between the price of crude oil and the prices of diesel fuel, gasoline, and liquefied petroleum gas, respectively over the period 1970 to 1997), and since reliable data on the prices of refined petroleum products paid by farmers do not go back to 1963, the crude oil price (refiner acquisition cost) is used as a suitable proxy.

[1988] and the Conservation Technology Information Center [annual]. Given the focus of the analysis is on the impact of the price of energy on the adoption of conservation tillage practices relative to the use of conventional tillage practices, the percentage of total planted acreage on which conservation tillage is practiced is used as the dependent variable in the analysis.

(C) COINTEGRATION TESTS

Table 9.2 reports Augmented Dickey-Fuller (ADF) and the Phillips-Perron (PP) tests conducted to investigate for autoregressive unit roots in the percent of planted acreage on which conservation tillage is used and the real price of crude oil.[62] The regressions include a constant and twelve lags of the dependent variable to adjust for higher-order autoregressive or mixed ARMA (autoregressive-moving average) processes. Schwert [1989] has shown that the ADF test performs adequately in Monte Carlo comparisons with tests proposed by Phillips [1987] and Phillips and Perron [1988] which correct for conditional heteroscedasticity and weak dependence of the cointegrating regression residuals. The test results find that the series are stationary and do not need to be differenced to induce stationarity. That is, the series under consideration are integrated of order 0.

Table 9.2. Computed ADF and PP Unit Root Tests for Determining the Order of Integration[1]

Series	ADF	PP
Percent of Planted Acres on which Conservation Tillage is Used	-1.98	-2.01
Real Price of Crude Oil	-1.66	-1.95

[1] The critical values for the augmented Dickey-Fuller (ADF) test statistic at the 5% and 1% levels of significance are -2.86 and -3.43, respectively. The critical values for the Phillips-Perron (PP) test statistic at the 5% and 1% levels of significance are -2.86 and -3.43, respectively.

[62] Note that these tests are one-sided.

Summary results for the static cointegrating regressions (i.e., relationship (3)) are presented in Table 9.3. Provided in the table are Cointegrating Regression Durbin-Watson statistics (as proposed by Sargan and Bhargava [1983]) as well as the t-statistics for both Augmented Dickey-Fuller and Phillips-Perron tests for stationarity of the regression residuals.[63,64] Although not reported, the OLS estimates of cointegrating constants were statistically significant and positive. In addition the Ljung-Box modified Q-statistic (Ljung and Box [1978]) for each regression was very large with marginal significance levels less than 0.001 percent in each case. (The same results were obtained when longer lag lengths (of 18 and 24) of the dependent variable were considered.)

Table 9.3. Augmented Dickey-Fuller and Phillips-Perron Tests for Static Cointegrating Regressions.

Series	CRDW[1]	ADF[2]	PP3
Xt/Yt [4]	0.7613	-4.67	-5.21

[1] CRDW denotes the cointegrating regression Durbin-Watson statistic. The critical values for the CRDW are 0.386 and 0.511 at the 5% and the 1% levels, respectively.
[2] ADF denotes the augmented Dickey-Fuller test statistic. For the Augmented Dickey-Fuller test, the critical value is -3.50 at the 10 % level.
[3] PP denotes the Phillips-Perron test statistic. For the Phillips test, the critical value is -3.04 at the 10 % level.
[4] Xt denotes the percent of total planted acres devoted to conservation tillage and Yt denotes the real price of energy.

The relatively large values for the CRDW, ADF and the PP statistics indicate the absence of a cointegrating relationship between the percent of planted

[63] Based on a Monte Carlo study, Engle and Granger [1987] recommend the ADF test. Both tests are used here, however, since the PP test is frequently used in applied econometric work and reporting the PP statistic affords the reader an opportunity to compare the two tests in detecting unit roots. Hall [1986], MacDonald and Murphy [1989], and Nachane and Chrissanthaki [1991], among others, employ both to test for stationarity in cointegrating regression residuals.
[64] Lags of length 12 were used in the estimation of the test statistics.

acreage on which conservation tillage is used and the real price of crude oil. That is, since the computed values of the test statistics exceed the critical values in all instances for both the Augmented Dickey-Fuller test and the Phillips-Perron test, the null hypothesis of a cointegrating relationship is rejected at the 10 percent level (Banerjee et al. [1993]).[65]

A closer examination of the results from the static cointegration tests, however, leads one to be concerned with the relatively large computed Q-statistic.[66] This value suggests the possibility that the inability to accept the null hypothesis of cointegration may be due to the low power of the test because of autocorrelated residuals.

Table 9.4. Johansen-Juselius Cointegration Test.

	Johansen Statistics[1]	
Series	Ho: $r \leq 1$	Ho: $r = 0$
Xt/Yt[2]	3.04	8.63

[1] If r denotes the number of significant cointegrating vectors, the Johansen test statistics test the null hypothesis of at most one and zero cointegrating vectors, respectively. The 5 percent critical value for Ho: $r \leq 1$ is 9.094 and for Ho: r=0, it is 20.168.
[2] Xt denotes the percent of total planted acres devoted to conservation tillage, Yt denotes the real price of energy.

Jenkinson [1986] notes the low power of the CRDW test as well as the ADF test when the residuals in static cointegrating regressions display stationarity but exhibit an autoregressive pattern. For example, the value of the CRDW statistic in Table 9.3 implies a first order autoregressive coefficient that is very close but not equal to one. Sargan and Bhargava [1983] find that the power of the CRDW test for random walk behavior (the null hypothesis)

[65] Actually, the null hypothesis is that the residual on the cointegrating regression is stationary (integrated of order zero). This translates into testing the null hypothesis of whether the two series under consideration are cointegrated.

[66] The computed Ljung-Box modified Q-statistic based on 32 degrees of freedom for the was 81.73. The critical value at the 5 percent level is 46.19.

against the alternative hypothesis that $u_t = \psi u_{t-1} + \xi_t$ becomes very low as ψ approaches one.[67]

Another approach to testing for cointegration that avoids the shortcomings of the Augmented Dickey-Fuller, the Phillips-Perron, and the Cointegrating Regression Durbin-Watson tests is that suggested by Johansen and Juselius [1990]. This is a maximum likelihood procedure involving the rank of the matrix composed of column vectors of the variables involved in the analysis. If the rank is zero, the variables are not cointegrated. If the rank is r, however, there exist r possible independent stationary linear combinations.

Table 9.4 reports the results of the cointegration test using the maximum likelihood approach of Johansen and Juselius. In the table, for r which denotes the number of significant cointegrating vectors, the Johansen statistics test the hypothesis of at most one and zero cointegrating vectors, respectively. A constant was included in the vector autoregressions. The null hypothesis of at most one cointegrating vector (Ho: $r \leq 1$) is not rejected at the 5 percent level while the null hypothesis of zero cointegrating vectors (Ho: $r = 0$) is rejected. These results are inconsistent with the results of the previous cointegration tests.

Table 9.5. Final Restricted Joint Error Correction Model Representations[1]

Series			
Xt =	0.1006 +	0.6981 Xt-1 -	0.0047 ECt-1
	(0.6721)	(0.2707)	(0.0021)
SEE = 26.55			
Q(32) = 21.39			

[1] Standard errors of the estimates are in parentheses. SEE denotes the standard error of the regression. Q is the computed Ljung-Box modified Q-statistic with the number of degrees of freedom in parentheses. EC denotes the error correction term.

One final test of cointegration involves the estimation of error correction models. Engle and Granger [1987] demonstrate that cointegration implies an

[67] Note that the ξ_t are assumed to be identically and independently distributed with a mean of zero and a finite variance.

error correction model. Table 9.5 presents the final, restricted error correction models. Included in the tables are OLS parameter estimates with standard errors of the estimates in parentheses, Ljung-Box modified Q-statistics, and the standard errors of the regressions, SEE.

The basic model-building strategy delineated by Granger and Weiss [1983] and Engle and Granger [1987] was followed in deriving the final parsimonious error correction models (ECMs) from the initial overparameterized specifications.[68] This procedure involves dropping all insignificant lagged values of both variables and imposing the restrictions implied by the lagged error correction term, ECt-1.[69] The initial error correction models were specified such that they included a constant and twelve lags of both dependent and independent variables.

The coefficient estimate on error correction term is statistically significantly different than zero. Thus, the results from the error correction model specification are consistent with those obtained from the maximum likelihood approach of Johansen and Juselius but not the other static cointegrating tests.

These results highlight the importance of using not just a static cointegration approach to investigate the relationship between the percent of planted acreage on which conservation tillage is used and the real price of crude oil. Clearly, dynamic tests should become an integral part of any such investigations. This especially true, as is the case in this study, when the residuals from the cointegrating regressions are stationary but highly autocorrelated.

Relying on the results of the error correction model estimation and the maximum likelihood approach of Johansen and Juselius, there is an indication that cointegration between the percent of planted acreage on which conservation tillage is used and the real price of crude oil. Moreover, these results are fairly robust when different lag lengths and time subperiods are considered.

---

[68] Note that it would be tempting to use data based criteria like the Akaike Information Criterion (AIC) (Akaike [1974]) for selecting the model specification. Such a procedure, however, is likely to result in reduced power of the test. See Engle and Granger [1987] for a discussion of this.

[69] The estimated slope coefficient of the static cointegrating regression has been shown to be a consistent estimator of κ (Stock [1987]). It is this value that used in computing EC_t.

Quantifying the Effect of the Price of Energy on the Adoption of Conservation Tillage

(A) Preliminary Analyses

Since the price of energy clearly does affect the adoption of conservation tillage in the United States, the next issue deals with the nature of this effect. To address this, conventional empirical techniques will be used first to identify any lags in the impacts and any anomalies in the data and/or estimates and subsequently, its empirical character.

The functional specification considered relates the adoption of conservation tillage and the real price of crude oil in the current and previous periods. The lengths of the lags on the explanatory variables are determined by a zero restrictions test (Judge et al. [1985]). The results of the test indicate that the real price of crude oil in just the current period affects the extent to which conservation tillage is used in production agriculture in the United States. Thus, the impact of the real price of crude oil on conservation tillage use is immediate with no detectable lag effects.

Before turning to estimating empirically the relationship between the adoption of conservation tillage and the real price of crude oil, one additional item needs to be addressed. It involves the presence of data outliers. It is not uncommon in empirical work to find that the results are very much influenced by a subset of the total observations used in the estimation. As a check on the possibility that coefficient estimates were inordinately influenced by such a subset, the preliminary estimates were subjected to the regression diagnostics of Belsley, Kuh and Welsch [1980]. The fact that a subset of the data can have a disproportionate influence on the estimated parameters is of concern because it is quite possible that coefficient estimates in the model are generated primarily by this subset of the data rather than by all of the data equally. Belsley, Kuh and Welsch identify four diagnostic techniques to help in isolating influential data points: RSTUDENT, HAT DIAGONAL, COVRATIO, and DFFITS. Each of these diagnostics is employed here.

Regression diagnostics were performed on the basic relationship. The regression diagnostics - RSTUDENT, HAT DIAGONAL, COVRATIO, and

DFFITS - indicated no outliers.[70] There were no observations that are beyond the cutoff points.

(B) EMPIRICAL ESTIMATES

One way to characterize the production innovation cycle discussed above is by a logistic curve (Waterson [1984]). Beginning with Griliches [1957], this functional representation has been used repeatedly with success. It will be relied upon here. One adjustment will be made, however, to the standard logistic specification. There is no reason to a priori suppose that the upper bound asymptotic value is static. Whether it is, is a testable hypothesis. In the current instance, since it has previously been shown that energy considerations affect the adoption of conservation tillage and, consistent with the preliminary empirical results presented above, this upper bound will be made a function of the price of energy in just the current period.[71]

Consequently, the specific functional representation for the adoption and diffusion of conservation tillage in production agriculture in the United States is given as

$$(6)\quad Xt = (c1\ Yt)/(1 + c2 \exp(-c3\ t)) + wt$$

where Xt denotes the percent of total planted acres devoted to conservation tillage,
Yt denotes the real price of energy, wt denotes a vector white noise sequence with mean zero and finite variance, t denotes the time period, and c1, c2, and c3 are coefficients to be estimated.

To estimate equation (6), classical least squares with an adjustment for serial correlation might be employed. This is not feasible, however, because of the relationship's nonlinear nature. The appropriate technique is maximum

[70] An observation is determined to be an outlier if two or more of the four regression diagnostics cutoff points are exceeded.

[71] In preliminary analyses, the precise nature of the impact of the real price of crude oil was examined. Its effect on both the extent of use (as measured by the upper bound asymptote) and the rate of adoption were investigated. These issues were studies sequentially. While the real price of crude oil clearly affects the extent of use of conservation tillage in the United States as indicated by the empirical results reported, there is no statistically significant effect on the rate of adoption.

likelihood estimation (Judge et al. [1985]) that adjusts for serial correlation. The data used in the estimation are those previously discussed.

The estimation results are

$$(7)\ Xt = \underset{(1.2152)}{4.1914}\ Yt\ /$$

$$(1 + \underset{(148.4903)}{357.2011} \exp \ (\underset{(0.0309)}{-0.1927}\ t))$$

with the logarithm of the likelihood function = -93.23.

The estimate of the serial correlation coefficient was not statistically significantly different from zero at the ninety five percent level in preliminary analyses. Consequently, there was no adjustment for serial correlation in the computing the final estimates.

The results are interesting. While the impact of a change in the real crude oil price is relatively small, it is statistically significant at the 95 percent level. Thus, a ten percent rise in the real price of crude oil will lead to a increase in the percentage of total planted acres devoted to conservation tillage of 0.4 percent, all other things given.

(C) Testing for Structural Stability

To complete the analysis, one final issue needs to be investigated. Namely, has the underlying structural relationship between the adoption of conservation tillage and the real price of crude oil changed over the estimation period? That is, for example, in light of the energy crises during the decade of the 1970s, is the adoption of conservation tillage more (or perhaps less) responsive to the real price of crude oil today than it was, say, prior to 1970? An investigation of this is the subject of what follows.

To analyze this, the stability of the estimated relationship must be studied. Stability is defined here in the statistical sense of the estimated coefficients on the explanatory variables remaining constant over time. A method of determining whether a regression relationship is constant over a given time period has been developed by Brown, Durbin and Evans [1975]. Essentially, this approach necessitates the computation of one-period prediction residuals, which are obtained by applying the regression estimated with r-1 observations to predict the r th observation using k explanatory variables (including the constant). The method is based on a test statistic, S(r), which equals the ratio of the sum of squared residuals of one period prediction from the k + 1 period to the r th period to the sum of the squared residuals of one period prediction from the k + 1 period to the T th period, where T denotes the sample size. The null hypothesis that the regression relationship is constant over time implies that the expected value of the test statistic S(r), E(S(r)), will lie along (in a statistical sense) its mean value line. For a more complete description of this test, the interested reader is referred to Harvey [1981].

The results[72] suggest that, for the estimated equation, the underlying structural relationship did not significantly change over the period 1963 to 1997. There are no years in which the test statistic ventures above or below the upper or lower bound, respectively, of the ninety five percent confidence interval. That is, the relationship between the adoption of conservation tillage and the real price of crude oil is stable over the entire sample period. The implication of these results is transparent. Events over the past three and one half decades have left virtually unchanged the impact that the real price of crude oil has had on the relative adoption of conservation tillage in the United States. One must be careful, however, in inferring that the adoption of conservation tillage remained constant over time. Clearly this is not so. The estimation results show that the real price of crude oil impacts the extent of adoption of conservation tillage. The magnitude of this, however, over the period 1963-1997 did not vary.

CONCLUSION

The relationship between energy and the adoption of conservation tillage is of special importance in addressing concerns about the impact of agricultural production on the environment in the United States. It has been the subject this paper. After establishing that a relationship exists between the price of energy and the adoption of conservation tillage using cointegration techniques, the relationship is quantified. It is shown that while the real price of crude oil, the proxy used for the price of energy, does not affect the rate of adoption of conservation tillage, it does impact the extent to which it is used. Finally, there is no structural instability in the relationship between the rela-

[72] Because they provide few insights, complete numerical results are not presented.

tive adoption of conservation tillage and the real price of crude oil over the period 1963 to 1997.

In a policy context, the results of the foregoing analysis show that the price of energy can be used to promote the adoption of conservation tillage. By increasing the real price of energy via, for example, a tax, the extent to which conservation tillage is used will expand. On the negative side, however, is the fact that a fairly substantial increase must be made in order for any significant change to occur in the adoption of conservation tillage.

REFERENCES

Akaike, A., "A New Look at Statistical Model Identification," *IEEE Transactions on Automatic Control*, Vol. 19 (1974), pp. 716-723.

Alchien, A., "Costs and Output," *The Allocation of Economic Resources*, Stanford University Press, Stanford, 1959

Baker, J., "Hydrologic Effects of Conservation Tillage and Their Importance to Water Quality," in T. Logan, J. Davidson, J. Baker, and M. Overcash (eds.), *Effects of Conservation Tillage on Groundwater Quality: Nitrates and Pesticides*, Lewis Publishers, Chelsea, MI, 1987, pp. 113-124.

Baker, J., and H. Johnson, "The Effect of Tillage Systems on Pesticides in Runoff from Small Watersheds," *Transactions of the American Society of Agricultural Engineers*, Vol. 22 (1979), pp. 554-559.

Banerjee, A., J. Dolado, J. Galbraith, and D. Hendry, *Co-Integration, Error Correction, and the Econometric Analysis of Non-Stationary Data*, Oxford University Press, Oxford, 1993.

Batte, M., "Technology and Its Impact on American Agriculture," *Size, Structure, and the Changing Face of American Agriculture*, Westview Press, Inc., Boulder, CO, 1993.

Batte, M., L. Forster, and K. Bacon, *Performance of Alternative Tillage Systems on Ohio Farms*, Department of Agricultural Economics and Rural Sociology, The Ohio State University, Columbus, OH, 1993.

Belsley, D., E. Kuh, and R. Welsch, *Regression Diagnostics*, John Wiley and Sons, Inc., New York, 1980.

Box, G.E.P., and G.M. Jenkins, *Time Series Analysis: Forecasting and Control*, Holden Day, Inc., San Francisco, 1971.

Brown, R.L., J. Durbin and J. Evans, "Techniques for Testing the Constancy of Regression Relationships Over Time," *Journal of the Royal Statistical Society*, Vol. 37 (1975), pp. 149-163.

Carter, M., and H. Kunelius, "Adapting Conservation Tillage in Cool, Humid Regions," *Journal of Soil and Water Conservation*, Vol. 45 (1990), pp. 454-456.

Conservation Technology Information Center, *National Crop Residue Management Survey*, West Lafayette, IN, annual.

Council of Economic Advisors, *Economic Report of the President*, U.S. Government Printing Office, Washington, 1979.

Dick, W.A., and T.C. Daniel, "Soil Chemical and Biological Properties as Affected by Conservation Tillage: Environmental Implications," in T. Logan, J. Davidson, J. Baker, and M. Overcash (eds.), *Effects of Conservation Tillage on Groundwater Quality: Nitrates and Pesticides*, Lewis Publishers, Chelsea, MI, 1987, pp. 315-339.

Energy Information Administration, Monthly Energy Review, DOE/EIA-0035 (95/01), U.S. Department of Energy, Washington, December 1997.

Engle, R., and C. Granger, "Co-integration and Error Correction: Representation, Estimation, and Testing," *Econometrica*, Vol. 55 (1987), pp. 251-276.

Engle, R., and B. Yoo, "Forecasting and Testing in Co-Integrated Systems," *Journal of Econometrics*, Vol. 35 (1987), pp. 143-160.

Fawcett, R.S., "Overview of Pest Management for Conservation Tillage Systems," in T. Logan, J. Davidson, J. Baker, and M. Overcash (eds.), *Effects of Conservation Tillage on Groundwater Quality: Nitrates and Pesticides*, Lewis Publishers, Chelsea, MI, 1987, pp. 6-19.

Frye, W., "Energy Requirements in No-tillage," in R.E. Phillips and S.H. Phillips (eds.), *No-Tillage Agriculture: Principles and Practices*, Van Nostrand Reinhold Company, New York, NY, 1984.

Frye, W., "Energy Use in Conservation Tillage," *Farming for a Better Environment, Soil and Water Conservation Society*, Ankeny, IA, 1995.

Geweke, J., "Inference and Causality in Economic Time Series Models," in *Handbook of Econometrics*, Z. Griliches and M. Intriligator (eds.), North-Holland Publishing Company, Amsterdam, 1984.

Geweke, J., R. Meese, and W. Dent, "Comparing Alternative Tests of Causality in Temporal Systems: Analytic Results and Experimental Evidence," *Journal of Econometrics*, Vol. 21 (1983), pp. 161-194.

Glenn, S., and J.S. Angle, "Atrazine and Simazine in Runoff from Conventional and No-Till Corn Watersheds," *Agriculture, Ecosystems, and Environment*, Vol. 18 (1987), pp. 273-280.

Granger, C., "Developments in the Study of Cointegrated Economic Variables," *Oxford Bulletin of Economics and Statistics*, Vol. 48 (1986), pp. 213-228.

Granger, C., "Some Recent Developments in the Concept of Causality," *Journal of Econometrics*, Vol. 39 (1988), pp. 199-211.

Granger, C., and A. Weiss, "Time Series Analysis of Error-Correction Models," *Studies in Econometrics, Time Series, and Multivariate Statistics*, S. Karlin, T. Amemiya, and L. Goodman (eds.), Academic Press, Inc., New York, 1983.

Granger, C., and P. Newbold, "Spurious Regressions in Econometrics," *Journal of Econometrics*, Vol. 2 (1977), pp. 111-20.

Granger, C., and A. Weiss, "Time Series Analysis of Error-Correction Models," *Studies in Econometrics, Time Series, and Multivariate Statistics*, S. Karlin, T. Amemiya, and L. Goodman (eds.), Academic Press, Inc., New York, 1983.

Gray, R., J. Taylor, W. Brown, "Economic Factors Contributing to the Adoption of Reduced Tillage Technologies in Central Saskatchewan," *Journal of Plant Science*, Vol. 37 (1996), pp. 7-17.

Griffith, D., J. Mannering, and C. Richey, "Energy Requirements and Areas of Adaptation for Eight Tillage-Planting Systems for Corn," in W. Lockertz (ed.), *Agriculture and Energy*, Academic Press, New York, NY, 1977.

Griliches, Z., "Hybrid Corn: An Exploration in the Economics of Technical Change," *Econometrica*, Vol. 25 (1957), pp. 501-522.

Hall, J., L. Hartwig, and L. Hoffman, "Cyanazine Losses in Runoff from No-Tillage Corn in "Living" and Dead Mulches vs. Unmulched, Conventional Tillage," *Journal of Environmental Quality*, Vol. 13 (1984), pp. 105-110.

Hall, S., "An Application of the Granger and Engle Two-Step Estimation Procedure to the United Kingdom Wage Data," *Oxford Bulletin of Economics and Statistics*, Vol. 48 (1986), pp. 229-239.

Harvey, A., The Econometric Analysis of Time Series, Phillip Allen, Ltd., Oxford, 1981.

Jenkinson, T., "Testing Neo-Classical Theories of Labor Demand: An Application of Cointegration Techniques," *Oxford Bulletin of Economics and Statistics*, Vol. 48 (1986), pp. 241-251.

Johansen, S., and K. Juselius, "Maximum Likelihood Estimation and Inference on Cointegration - With Applications to the Demand for Money," *Oxford Bulletin of Economics and Statistics*, Vol. 52 (1990), pp. 169-210.

Judge, G., W. Griffiths, R.C. Hill, H. Lutkepohl, and T.C. Lee, *The Theory and Practice of Econometrics*, 2nd ed., John Wiley and Sons, Inc., New York, 1985.

Kislev, Y., and N. Schori-Barach, "The Process of an Innovation Cycle," *American Journal of Agricultural Economics*, Vol. 55 (1973), pp. 28-37.

Ljung, G., and G. Box, "On a Measure of Lack of Fit in Time Series Models," *Biometrika*, Vol. 66 (1978), pp. 297-304.

MacDonald, R., and P. Murphy, "Testing for the Long Run Relationship between Nominal Interest Rates and Inflation Using Cointegration Techniques," *Applied Economics*, Vol. 21 (1989), pp. 439-447.

Mikesell, C., J. Williams, and J. Long, "Evaluation of Net Return Distributions from Alternative Tillage Systems for Grain Sorghum and Soybean Rotations," *North Central Journal of Agricultural Economics*, Vol. 10 (1988), pp. 255-271.

Mills, T., *Time Series Techniques for Economists*, Cambridge University Press, Cambridge, 1990.

Nachane, D., and A. Chrissanthaki, "Purchasing Power Parity in the Short and Long Run: A Reappraisal of the Post-1973 Evidence," *Applied Economics*, Vol. 23 (1991), pp. 1257-1268.

National Research Council, Soil and Water Quality, National Academy Press, Washington, DC, 1993.

Nowak, P., "Why Farmers Adopt Production Technology," *Journal of Soil and Water Conservation*, Vol. 47 (1992), pp. 14-16.

Nowak, P., and G. O'Keefe, "Evaluation of Producer Involvement in the United States Department of Agriculture 1990 Water Quality Demonstration Projects," Baseline Report submitted to the *U.S. Department of Agriculture*, Washington, DC, November 1992.

Nowak, P., and G. O'Keefe, "Farmers and Water Quality: Local Answers to Local Issues," Draft Report submitted to the *U.S. Department of Agriculture*, Washington, DC, September 1995.

Onstad, C., and W. Voorhees, "Hydrologic Soil Parameters Affected by Tillage," in T. Logan, J. Davidson, J. Baker, and M. Overcash (eds.), *Effects of Conservation Tillage on Groundwater Quality: Nitrates and Pesticides*, Lewis Publishers, Chelsea, MI, 1987, pp. 274-291.

Pagoulatos, A., D. Debertin, and F. Sjarkowi, "Soil Erosion, Intertemporal Profit, and the Soil Conservation Decision," *Southern Journal of Agricultural Economics*, Vol. 21 (1989), pp. 55-62.

Phillips, P., "Time Series Regression with a Unit Root," *Econometrica*, Vol. 55 (1987), 277-301.

Phillips, P., and S. Ouliaris, "Asymptotic Properties of Residual Based Tests for Cointegration," *Econometrica*, Vol. 58 (1990), pp. 165-193.

Phillips, P., and P. Perron, "Testing for a Unit Root in Time Series Regression," *Biometrika*, Vol. 75 (1988), pp. 335-346.

Pierce, D.A., and "Forecasting in Dynamic Models with Stochastic Regressors," *Journal of Econometrics*, Vol. 3 (1975), pp. 349-374.

Pierce, D.A., and L.D. Haugh, "Causality in Temporal Systems," *Journal of Econometrics*, Vol. 5 (1977), pp. 265-293.

Pierce, D., and L. Haugh, "The Characterization of Instantaneous Causality," *Journal of Econometrics*, Vol. 7 (1979), pp. 257-259.

Priestley, M.B., *Spectral Analysis and Time Series*, Academic Press, London, 1981.

Sander, K., W. Witt, and M. Barrett, "Movement of Triazine Herbicides in Conventional and Conservation Tillage Systems," in D.L. Weigmann (ed.), *Pesticides in Terrestrial and Aquatic Environments*, Virginia Water Resources Center, Virginia Polytechnic Institute and State University, Blacksburg, pp. 378-382.

Sargan, J., and A. Bhargava, "Testing Residuals from Least Squares Regression for Being Generated by the Gaussian Random Walk," *Econometrica*, Vol. 51 (1983), pp. 153-174.

Sawa, T., "Information Criteria for Discriminating Among Alternative Models," *Econometrica*, Vol. 46 (1978), pp. 1273-1291.

Schertz, D., "Conservation Tillage: An Analysis of Acreage Projections in the United States," *Journal of Soil and Water Conservation*, Vol. 33 (1988), pp. 256-258.

Schwert, G., "Test for Unit Roots: A Monte Carlo Investigation," *Journal of Business and Economic Statistics*, Vol. 7 (1989), pp. 147-160.

Stock, J., "Asymptotic Properties of Least Squares Estimators of Cointegrating Vectors," *Econometrica*, Vol. 55 (1987), pp. 1035-1056.

Torgerson, D., J. Duncan, and A. Dargan, *Energy and U.S. Agriculture*, U.S. Department of Agriculture, Economic Research Service, Washington, DC, 1987.

U.S. Congress, *Public Law No. 96-294*, 94 Stat. 611, 16 USC 590h (1980).

U.S. Environmental Protection Agency, Office of Water, *National Water Quality Inventory - 1994 Report to Congress*, Office of Water, Washington, DC, 1995.

Uri, N., "Conservation Tillage and Input Use," *Environmental Geology*, Vol. 29 (1997), pp. 188-200.

Uri, N., "The Role of Public Policy in the Use of Conservation Tillage in U.S. Agriculture," *International Journal of Energy, Environment and Economics*, forthcoming 1998.

Uri, N., and K. Day, "Energy Efficiency, Technological Change and the Dieselization of American Agriculture in the United States," Transportation Planning and Technology, Vol. 16 (1992), pp. 221-231.

Wagnet, R.J., "Processes Influencing Pesticide Loss with Water under Conservation Tillage," in T. Logan, J. Davidson, J. Baker, and M. Overcash (eds.), Effects of Conservation Tillage on Groundwater Quality: Nitrates and Pesticides, Lewis Publishers, Chelsea, MI, 1987, pp. 189-200.

Waterson, M., Economic Theory of Industry, Cambridge University Press, Cambridge, 1984.

Westra, J., and K. Olson, *Farmers' Decision Processes and the Adoption of Conservation Tillage*, Department of Applied Economics, University of Minnesota, Minneapolis, 1997.

Williams, J., R. Llewelyn, L. Goss, and J. Long, Analysis of Net Returns to Conservation Tillage from Corn and Soybeans in Northeast Kansas, *Kansas Agricultural Experiment Station Bulletin*, Kansas State University, Manhattan, 1989.

CHAPTER 10

THE ENVIRONMENTAL CONSEQUENCES OF THE CONSERVATION TILLAGE ADOPTION DECISION IN AGRICULTURE IN THE UNITED STATES

INTRODUCTION

Conservation tillage practices have been adopted by farmers in the United States over the past several decades aimed at maintaining or enhancing soil characteristics (Bruce et al. [1995), Ismail et al. [1994], Reicosky et al. [1995], Thorne and Thorne [1979] and Wood et al. [1991]). The majority of these practices have focused on affecting soil characteristics such as structure, organic matter content, and soil microbial populations that slow soil erosion and influence the movement of water in and through the soil (Office of Technology Assessment [1990]). Field studies conducted under natural rainfall on highly erodible land (14 percent slope) have compared erosion rates between tillage systems (Baker and Johnson [1979], Glenn and Angle [1987], Hall et al. [1984], and Sander et al. [1989]). Compared with moldboard plowing, no till generally reduced soil erosion by more than 90 percent while mulch tillage and ridge tillage reduced soil erosion by 50 percent or more.

Conservation tillage practices have a beneficial environmental impact. Yet, there is a lingering question as to their overall efficacy in reducing the impact of agricultural production on the environment. This is because the nature of the changes in agricultural chemical use required by conservation tillage have not yet been comprehensively investigated (Kellogg et al. [1994], National Research Council [1989, 1993]). What is known is that conventional tillage, in contrast to conservation tillage, contributes to pest control by destroying some perennial weeds, disrupting the life cycle of some insect pests, and burying disease inoculum (Holland and Coleman [1987]). Additionally, conventional tillage creates more bacterial activity and has a "boom-and-bust" effect on nutrient cycling processes while no till and other conservation tillage provides a slower but more even rate of nutrient release (Heichel [1987]). Thus, while pesticide use should increase under conservation tillage, what has not yet been determined conclusively is the extent to which it will increase and the environmental implications of this increase. Furthermore, the impact of conservation tillage on fertilizer use is uncertain and needs to be quantified. A study of these issues follows.

BACKGROUND

Modeling the effects of conservation tillage on the use of agricultural chemicals is a complex process. Conventional utility theory of the Von Neumann and Morgenstern [1947] type assumes that an individual producer (farmer) makes decisions under certainty or under certainty equivalence (Baumol [1972]), with a single objective such as maximizing net farm income. Farmers now, however, are required to take into account aspects other than profit or net return above variable costs. For example, a farmer must also consider the externalities associated with his or her farming operation such as soil erosion and surface water and groundwater contamination. In this analysis it is assumed that a farmer is maximizer of a multiattribute utility function. Multiattribute analysis offers a framework that permits a farmer to select among choices with different economic and environmental attributes. The complete analytics of this approach are not developed here. The interested reader is referred to Keeney and Raiffa [1976].

The multiattribute utility approach assumes that each farmer is a potential adopter of each different tillage practice. For conservation tillage practices, however, this may not be true. That is, irrespective of the potential economic and environmental impacts, conservation tillage might not be an appropriate cropping practice due to, for example, site specific physical characteristics such as the slope of the cropland, the type of soil (texture and structure), etc. (Thorne and Thorne [1979]). Pudney [1989] proposes modeling this sort of situation using discrete random preference regimes. This approach assumes that a farmer adopting conservation tillage has a different preference structure than a farmer not adopting such a practice. Thus, a zero observation reflects the decision not to use conservation tillage. Consequently, in the first stage of the model a farmer decides whether or not he or she will adopt conservation tillage. Nonadopters are then dropped from the sample.

The second stage of the model focuses on conservation tillage adopters. First, it is assumed that some farmers cannot be induced to adopt conservation tillage irrespective of its impact on net farm income or the environment. Again, Pudney's discrete random preference regimes is plausible, i.e., where conservation tillage adopters have a different preference structure than nonadopters. In this case, zero observations (farmer does not adopt conservation tillage) reflect the decision not to employ a specific conservation tillage practice and only adopters determine the underlying structure of the use of conservation tillage. A Heckman model, described subsequently, is an appropriate statistical model for implementing this theoretical approach (Heckman [1979]).

Alternatively, one can envision a farmer evaluating his or her utility functions with and without the adoption of conservation tillage and then determine whether or not to adopt such a practice. This sequence is plausible if certain factors, such as environmental awareness and/or concern among different farmers and/or site specific physical characteristics, relate directly to the quantitative distinction between adoption and nonadoption of conservation tillage and are independent of the extent to which conservation tillage is adopted by a farmer. One can model this situation by first assuming that an individual farmer's utility function takes the form

$$(1)\ U = U(\pi(Py, x1, x2, ..., xk \mid \alpha\tau), Q(z1, z2, ..., zn))$$

where π denotes the profit function, y denotes the production function, P denotes the price of output which is assumed to be given in the competitive market for the agricultural commodity,[73] x1, x2, ..., xk denote the factors of production (inputs) used in the production of the commodity, τ denotes the adoption (use) of a conservation tillage practice, α is equal to one if a conservation tillage practice is adopted or has the potential to be adopted and zero other-

[73] For ease of exposition, the production of just a single commodity is being considered. Extension of the analysis to the production of multiple commodities is straightforward (Zilberman and Marra [1993]).

wise, and Q (z1, z2, ..., zn) represents the characteristics of the cropland and the farmer that serve to potentially influence the adoption of a conservation tillage practice. Note that $U(\pi, Q)$ is a monotonically increasing and concave Von Neumann-Morgenstern utility function.

The profit function is given as

$$(2)\ \pi = P\,y(x^1, x^2, ..., x^k | \alpha\tau) - c^1x1 - c^2x^2 - ... - c^kx^k - \Omega$$

where c^1, c^2, ..., ck denote the stochastic imputed price of the factors of production, and Ω denotes the fixed cost of production. For the production function, $\partial y/\partial x^k \geq 0$ and $\partial^2 y/\partial^2 x^k \leq 0$ (Anderson et al. [1977]).

Assuming the farmer has as his or her objective the maximization of expected utility which is a function of profits and conditional upon the adoption of conservation tillage and the characteristics of the cropland and the farmer, the problem becomes

(3) maximize $E(U(\pi(Py, x1, x2, ..., xk | \alpha\tau), Q(z1, z2, ..., zn))$
where $\partial U/\partial xi > 0$ and $\partial^2 U/\partial^2 xi < 0$.[74]

Substituting (2) into (3), the expected utility is

$$(4)\ E(U(Py(x1,x2,...,xk | \alpha\tau) - c1x1 - c2x2 - ... - ckxk - \Omega), Q(z1,z2,...,zn)).$$

The first order conditions for the maximization problem (3) with regard to a specific factor of production, xi, are

$$(5)\ E\,((\partial U(\pi, Q)/\partial xi)(P(\partial y(\alpha\tau)/\partial xi) - ci) = 0.$$

Since $\partial U(\pi, Q)/\partial xi > 0$ by assumption, the optimal use of factor of production xi occurs when

[74] These inequalities have economic interpretations. The first, $\partial U/\partial x_i > 0$, indicates that an increase in the use of factor of production x_i will increase output y which in turn will increase profit and hence utility. The second inequality, $\partial^2 U/\partial^2 x_i < 0$, used in conjunction with the first inequality indicates that while the increase in the use of a factor of production will increase output and hence profit and utility, the increase will be at a decreasing rate. That is, for a given factor of production, the production of the agricultural commodity and therefore profit and utility is characterized by diminishing marginal returns (Stigler [1966]).

(6) $P(\partial y(\alpha\tau)/\partial xi) = ci.$

That is, the factor of production will be used up to the point where the value of the marginal product associated with that factor of production is just equal to its cost. This is a standard result from neoclassical microeconomic theory (Stigler [1966]). The optimal use of the factor of production, however, will be conditioned by the tillage practice used because the production function is dependent on whether conventional tillage or conservation tillage is employed. Therefore, it is not possible to conclude a priori precisely what impact conservation tillage will have on input usage as the farmer endeavors to maximize profit-based expected utility that reflects the environmental externalities associated with agricultural production.

In the context of an explicit adoption decision, a farmer will compare his or her utility at zero adoption with the utility at the level of conservation tillage adoption if it is decided to use one of these practices. The criterion for adoption is

$$(7)\quad A = \begin{cases} 1 \text{ if } \zeta > 0 \\ 0 \text{ otherwise} \end{cases}$$

$$\zeta = U(\pi^*, Q^*) - U(\pi, Q)$$

where $U(\pi^*, Q^*)$ denotes the utility associated with the adoption of conservation tillage and $U(\pi, Q)$ denotes the utility associated with nonadoption.[75] For the farmer who will not adopt conservation tillage under any circumstances, the indifference curves between conservation tillage adoption and nonadoption must be upward sloping and $U(\pi^*, Q^*) - U(\pi, Q)$ will be negative. Double hurdle (or Cragg) models are statistical counterparts of this behavioral structure.

[75] For ease of exposition just a single conservation tillage practice is being compared to conventional tillage. Obviously, the analysis can be expanded to include the comparison of the various conservation tillage practices and conventional tillage as well as a comparison among conservation tillage practices themselves.

STATISTICAL MODELS

The first model endeavors to associate the use of inputs, specific cropland characteristics, and a farmer's characteristics with the decision to adopt conservation tillage. A probit model is used for this purpose. For all remaining models, the nonadopters are deleted from the sample. This implies that nonadopters are not at a standard corner solution. That is, changes in such things as the price of output and the prices of the factor inputs will not induce them to adopt conservation tillage.

Attention is now focused on the econometrics of modeling the conservation tillage adoption decision and the extent of adoption decision. First, for ease of exposition, it is assumed that both the adoption and extent of adoption equations are linear in parameters (β, γ) with additive disturbance terms e and u, and the matrices V and W contain variables hypothesized to influence the adoption and extent of adoption decisions, respectively.

Mathematically, the conservation tillage adoption specification is

(8) $\zeta = \beta'V + e$ for $\alpha = 1$ if $\zeta > 0$ and $\alpha = 0$ otherwise. Also, $e \sim n(0,1)$ by assumption.[76]

The extent of conservation tillage adoption specification is

(9a) $\tau = \alpha\tau^{**}$

where

$$\text{(9b)} \qquad \tau^{**} = \begin{cases} 0 \text{ if } \tau^* \leq 0 \\ \tau^* \text{ if } \tau^* > 0 \end{cases}$$

and

$$\tau^* = \gamma'W + u.$$

Note that it is assumed that $u \sim n(0,1)$.[77]

A positive adoption of conservation tillage τ is observed only if $\alpha = 1$ and $\tau^{**} > 0$. Cragg or double-hurdle models postulate that to observe positive adoption, the farmer must pass two hurdles: (1) be a potential adopter of con-

[76] That is, e is assumed to be normally distributed with mean 0 and standard deviation 1.

[77] That is, u is assumed to be normally distributed with mean 0 and standard deviation 1.

servation tillage and (2) actually adopt conservation tillage (Blundell and Meghir [1987], Cragg [1971], and Lee and Maddala [1985]). This allows for the possibility that zero conservation tillage adoption is a result of the extent of conservation tillage adoption decision. Hence, potential conservation tillage adopters may in fact not adopt any conservation tillage practice.

Assuming correlated equation error terms allows for the possibility that the conservation tillage adoption and the extent of adoption decisions are made simultaneously:

(10) $(e, u) \sim n(0, \Gamma)$

where

$$\Gamma = \begin{vmatrix} 1 & \sigma\rho \\ \sigma\rho & \sigma 2 \end{vmatrix}$$

where $n(0, \Gamma)$ denotes a bivariate normal distribution with mean 0 and standard deviation Γ, ρ denotes the correlation coefficient, and σ denotes the standard error.

Using 0 to denote zero adoption of conservation tillage and + to denote positive adoption, the likelihood function for the dependent Cragg model is

(11) $$\prod_{0}[1 - p(e > -\beta'V)\, p(u > -\gamma'W \mid e > -\beta'V)]$$

$$\prod_{+} [p(e > -\beta'V)\, p(u > -\gamma'W \mid -\beta'V)\, g(\tau^* \mid u > -\gamma'W, e > -\beta'V)]$$

or

(12) $$\prod_{0}[1 - \Phi(\beta'V, \gamma'W/\sigma, \rho)]$$

$$\prod_{+} [\Phi(\beta'V) + \sigma/\rho\, (\tau^* - \gamma'W) / \sqrt{1 - \rho}](1/\sigma)(\phi((\tau^* - \gamma'W)/\sigma))$$

where p denotes the probability, Φ and ϕ denote distribution and density functions respectively, and $g(\circ) = \Phi(\circ)/\phi(\circ)$.

If the error terms e and u are independent (i.e., $\rho = 0$), the independent Cragg model is obtained. This model assumes a feedback effect from the extent of conservation tillage adoption to the adoption decision (Deaton and Irish [1986]). The Tobit model (Maddala [1983]) is a nested version of the independent Cragg model with $\Phi(\beta'V) = 1$. One advantage of the Cragg over the Tobit model is that the former allows variables to have differing effects on the adoption and extent of adoption decisions.

A Heckman model assumes error terms of the adoption and extent of adoption equations are correlated and the adoption of conservation tillage decision dominates the extent of adoption decision. Domination implies that zero conservation tillage adoption is a result of the conservation tillage adoption decision and not the extent of conservation tillage adoption decision. Hence, only farmers with positive conservation tillage adoption levels are included in the extent of conservation tillage adoption decision. The model assumes the probability of a positive extent of conservation tillage adoption is equal to 1 given that $\alpha = 1$ or $p(\tau^* > 0 \mid \alpha = 1)$ and $g(\tau^* \mid \tau^* > 0, \alpha = 1) = g(\tau^* \mid \alpha = 1)$. The likelihood function corresponding to this model is

$$(13) \qquad \prod_{0}[1 - p(e > -\beta'V)] \prod_{+} [(p(e > -\beta'V)] (g (\tau^* \mid u > -\gamma'W))$$

with the log of the likelihood function written as

$$(14) \qquad \prod_{+} \{-(1/2) \ln \sigma^2 + \ln (\phi((\tau^* - \gamma'W)/\sigma))$$

$$+ \rho J((\tau^* - \gamma'W)/\sigma)/(\sqrt{1 - \rho})\}$$

$$+ \quad \sum_{0} \ln (1 - \Phi(\beta'V))$$

where $J = \Phi$-1 (G) and G denotes the distribution function for u (Lee and Maddala [1985]).

In contrast to the Cragg models, the Heckman model assumes that a farmer with no conservation tillage adoption provides no restrictions on the parameters of the extent of conservation tillage adoption equation. To see this, note that $p(u > -\gamma' W \mid e > -\beta' V)$ does not appear in the likelihood function for the Heckman model or in the expectation of τ^*, denoted by g, conditional upon $e > -\beta' V$.

The Heckman model is simplified if the conservation tillage adoption and the extent of adoption equations are independent (i.e., $\rho = 0$). This model, termed the complete dominance model, separates into two independent components: (1) a probit model for the adoption relationship and (2) an ordinary least squares specification for the extent of conservation tillage adoption relationship using observations only on farmers who adopt conservation tillage (Maddala [1983] and Heckman [1979]).

Data and Model Specification

The basic data used in estimating the conservation tillage adoption and extent of conservation tillage adoption relationships are for corn farms in the United States for 1987. The data come from the 1987 Farm Costs and Returns Survey (FCRS) conducted in February and March 1988 by the National Agricultural Statistics Service of the U.S. Department of Agriculture. The FCRS is a stratified, multiframe survey consisting of a list frame and an area frame. The list frame farms were stratified by economic size, while area frame farms were stratified by use type. The survey is a full probability survey with all producers having a likelihood of being selected in the sample. Multiple versions of the FCRS are integrated into a single survey to obtain data simultaneously on farm organization, farm income, and expenses, assets and debt, and operator and household characteristics. Commodity-specific versions of the FCRS, which are conducted on a four-year rotation, obtain data on enterprise production practices used in cost-of-production estimation. Data from all versions of the 1987 FCRS were used because corn was one of the commodities surveyed during that year. The sample of corn farms consisted of 1222 observations of which 825 were usable. Observations were deleted from the sample either because relevant information (for estimation purposes) was omitted or the data were seemingly incorrect.[78] The data on input use are in terms of ex-

[78] Incorrect data consisted of negative expenditures for factor inputs, acres treated with pesticides and on which fertilizer was applied being in excess of the number of acres planted, the

penditures and not physical quantities. The survey does not collect data on quantities of inputs used or on input prices. (Combining expenditure data with price data would allow for the computation of quantity data.) Moreover, the survey that does collect data on input prices is problematic (Uri 1994). Thus, quantity data are simply not available.[79]

Merged with the FCRS data were specific topography and soil texture data taken from the National Resource Inventory (Soil Conservation Service [1982]) which contains county-level data. For each county, the National Resource Inventory (NRI) sampled the physical characteristics of all non-federal rural land at several randomly selected points. Within county observations on soil texture and slope are quantified and averaged. The average includes only cropland observations. The NRI data are matched to a specific farm based on the county in which the farm is located.

The topography variable, soil slope, measures the average cropland slope (in percent) for the county.

Table 10.1. Variable Definitions

Variable	Definition
FEXP	Per acre[1] expenditures on fertilizer and soil conditioners
PEXP	Per acre expenditures on insecticides, herbicides, fungicides, nematicides, and defoliants
FUEL	Per acre expenditures on energy including diesel fuel, gasoline, liquefied petroleum gas, and electricity
LABOR	Per acre expenditures on hired labor
OWNLABOR	Number of hours per acre of owner/operator labor devoted to corn production
IRRIGATE	Proportion of planted acres irrigated
WATER	Per acre quantity of water applied on irrigated acreage (in acre/feet)
SEED	Per acre seeding rate

number of irrigated acres being greater than the total number of acres planted, the seeding rate being unreasonably high (in excess of 30,000 per acre), etc.

[79] It is not clear what impact, if any, this sort of measurement error has on the empirical results (Uri [1994]).

Table 10.1. Variable Definitions (continued)

NOPEST	Proportion of acres that received no pesticide treatment
YIELD	Per acre corn yield (bushels)
ARP	The proportion of acres idled under the acreage reductionprogram[2]
TOTALACRES	The total number of acres operated by the farmer/operator
FARMTYPE	Defined to equal one if the farm is classified as a cash grain enterprise and zero otherwise (occurring, for example, if the farm is primarily operated to produce dairy products or beef or hogs)
OWNTYPE1[3]	Defined to equal one if the farm is operated by a single owner and zero otherwise
OWNTYPE2	Defined to equal one if the farm is operated as a partnership and zero otherwise
EDUCATION1	Defined to equal one if the farmer/operator graduated from college and zero otherwise
EDUCATION2	Defined to equal one if the farmer/operator has some college but did not graduate and zero otherwise
EDUCATION3	Defined to equal one if the farmer/operator has only a high school education and zero otherwise
AGE1	Defined to equal one if the farmer/operator is less than or equal to 30 years of age and zero otherwise
AGE2	Defined to equal one if the farmer/operator is more than 30 but less than or equal to 40 years of age and zero otherwise
AGE3	Defined to equal one if the farmer/operator is more than 40 but less than or equal to 50 years of age and zero otherwise
AGE4[4]	Defined to equal one if the farmer/operator is more that 50 but less than or equal to 60 years of age and zero otherwise
SLOPE	Average cropland slope (in percent)
TEXTURE1	Defined to equal one if the texture is less than or equal to 2.3 and zero otherwise (see text for a discussion)
TEXTURE2	Defined to equal one if the texture is greater than or equal to 3.6 and zero otherwise (see text for a discussion)
AVGYIELD	Average per acre corn yield for all farms in the county (bushels)
AVGTEMP	Average monthly temperature for June, July, and August (degrees Fahrenheit)
AVGRAIN	Average monthly rainfall for June, July, and August (inches)
PNT	Proportion of the total acreage planted to corn on which no tillage has been adopted
PMT	Proportion of the total acreage planted to corn on which mulch tillage has been adopted

1 This is per acre of corn planted.

2 See Lipton and Pollack [1989] for a technical definition of the acreage reduction program.

3 Note that the variables associated with categorical data are not all inclusive. That is, a variable that corresponds to the observations not found in the enumerated variable categories is not defined. This is done to avoid the problem of singularity in the estimation. Thus, for example, no variable is defined for a farm ownership type (such as a corporation) other than the one for a single owner and the one for a partnership.

4 The final category would be farmers who are more than 60 years of age.

Observations on soil texture are classified on a five point scale where 1 = sand, 2 = sandy loam, 3 = loam, 4 = clay loam, and 5 = clay. The numerical average for a county is then classified into one of three categories: sandy soil (texture < 2.3), loamy soil (2.3 < texture < 3.6), and clayey soil (texture > 3.6). The sand and clay variables capture the soil texture effect relative to loam.

Table 10.2. Sample means and standard deviations (in parentheses)

Variable	Full Sample	Conventional Tillage	No Tillage	Mulch Tillage
FEXP	$12.70	$12.03	$13.93	$13.37
	($7.83)	($7.94)	($7.85)	($7.50)
PEXP	$5.08	$4.73	$5.56	$5.72
	($3.34)	($3.20)	($3.36)	($3.62)
FUEL	$9.84	$10.06	$9.46	$9.34
	($7.17)	($7.37)	($7.09)	($6.41)
LABOR	$3.99	$3.85	$4.06	$4.08
	($5.51)	($5.03)	($5.53)	($5.49)
OWNLABOR	2.14	2.29	1.56	2.13
	(2.64)	(2.96)	(1.40)	(2.37)
IRRIGATE	0.03	0.03	0.02	0.03
	(0.10)	(0.10)	(0.08)	(0.10)
WATER	0.43	0.41	0.16	0.69
	(4.42)	(4.17)	(0.85)	(5.96)
SEED	23601	23440	23843	23746
	(3416)	(3592)	(2960)	(3272)
NOPEST	0.12	0.13	0.11	0.12
	(0.04)	(0.04)	(0.03)	(0.04)
YIELD	111.51	110.34	110.36	115.45
	(31.54)	(32.31)	(31.83)	(28.85)
ARP	0.11	0.11	0.12	0.13
	(0.08)	(0.09)	(0.08)	(0.08)
TOTALACRES	751.61	728.16	848.54	767.59
	(830.41)	(925.06)	(713.98)	(688.31)
FARMTYPE	0.92	0.91	0.94	0.95
	(0.05)	(0.06)	(0.07)	(0.08)
OWNTYPE1	0.77	0.78	0.73	0.80
	(0.41)	(0.41)	(0.45)	(0.40)

Table 10.2. Sample means and standard deviations (in parentheses) continued

OWNTYPE2	0.16	0.17	0.21	0.13
	(0.37)	(0.37)	(0.41)	(0.34)
EDUCATION1	0.12	0.10	0.18	0.12
	(0.32)	(0.30)	(0.39)	(0.32)
EDUCATION2	0.21	0.19	0.24	0.29
	(0.41)	(0.39)	(0.43)	(0.46)
EDUCATION3	0.47	0.49	0.44	0.46
	(0.49)	(0.50)	(0.50)	(0.50)
AGE1	0.07	0.08	0.10	0.08
	(0.27)	(0.27)	(0.30)	(0.28)
AGE2	0.74	0.75	0.79	0.81
	(0.25)	(0.27)	(0.26)	(0.25)
AGE3	0.07	0.08	0.05	0.06
	(0.21)	(0.24)	(0.23)	(0.25)
AGE4	0.09	0.06	0.04	0.04
	(0.19)	(0.20)	(0.22)	(0.22)
SLOPE	3.85	3.58	4.39	4.28
	(2.76)	(2.67)	(2.99)	(2.76)
TEXTURE1	0.11	0.11	0.14	0.08
	(0.30)	(0.31)	(0.35)	(0.28)
TEXTURE2	0.13	0.13	0.07	0.16
	(0.33)	(0.34)	(0.26)	(0.36)
AVGYIELD	111.97	110.36	112.03	112.56
	(31.93)	(30.15)	(31.67)	(29.75)
AVGTEMP	72.19	72.10	73.13	71.79
	(3.29)	(3.67)	(2.95)	(2.23)
AVGRAIN	4.04	4.01	4.56	3.73
	(1.41)	(1.53)	(1.34)	(1.04)
PNT	0.04	0.00	0.19	0.02
	(0.10)	(0.17)	(0.07)	
PMT	0.11	0.00	0.07	0.42
	(0.23)	(0.17)	(0.26)	
Number of Observations	825	481	159	218

Source: Data for variables FEXP through AGE4 and PNT and PMT were taken directly from or computed based on data taken from the 1987 Farm Costs and Returns Survey, data for variables SLOPE through TEXTURE2 were taken from the National Resource Inventory, AVGYIELD data were obtained from the National Agricultural Database, and AVGTEMP and AVGRAIN data were obtained from the National Oceanic and Atmospheric Administration. The text has relevant source citations.

To reflect the productivity of the soil, the average corn yield for the county in which the farm is located for 1987 is used. Soil productivity has not been shown conclusively to impact the use of conservation tillage. Some soils with high erosion potential are relatively more productive in terms of average yields (Webb et al. 1986]). But, whether farms possessing more productive soils on average typically use conservation tillage more extensively remains an open question (Diebel et al. [1993]). The yield data are taken from the National Agricultural Database compiled by the Economic Research Service of the U.S. Department of Agriculture.

Finally, weather data including mean daily temperature averaged over June, July, and August and total monthly rainfall averaged over June, July, and August were collected from the National Oceanic and Atmospheric Administration (NOAA). The data record for a specific farm was matched to the weather observation site nearest (in a spatial sense) the farm. Note that 1987 was not an abnormal weather year. The mean average daily temperature and rainfall for 1987 were not significantly different[80] from their historical averages.

Two different types of conservation tillage are considered - no till and mulch tillage.[81] For the usable sample, 19 percent (159 of the 825 farms in the sample) of the farms used no till, 26 percent (218) used mulch tillage, while 4 percent (33) used both no till and mulch tillage on some portion of the cropland used for corn production.

Presented in Table 10.1 are the variables and their definitions as used in the empirical models. Table 10.2 contains variable means and standard errors for the entire sample and for subsamples containing farmers who use only conventional tillage and farmers who have adopted conservation tillage (no till and mulch tillage) on some portion of their cropland. Note that commonly used statistical methods for computing the mean and standard deviation are inappropriate in this study since the data were obtained from a complex survey (Chamberlain [1986]). Unlike simple random sampling, and as observed previously, the selection of an individual farm in the FCRS is not equally likely for all farms included in the list frame. Some farms have a higher probability of selection than others. Differences in the probability of selection in-

[80] That is, they did not depart by more than two standard deviations from their historical mean levels.

[81] The number of observations on farms that adopted ridge till in the sample - 19 - was too small to get robust estimates of the parameters of the various models.

troduces a bias to the conventional estimates of the mean and standard deviation (Lee et al. [1989]). In order to overcome this problem, a weighted procedure in which the weights are equal to the inverse of the probability of selection must be used (Fuller [1984] and Fuller and Hidiroglou [1978]).

Hypothesizing that a given variable is interrelated with the conservation tillage adoption decision and not the extent of adoption decision or vice versa is difficult. Consequently, both conservation tillage adoption and the extent of conservation tillage adoption are postulated to be functions of the various FCRS variables including expenditures on and use of factor inputs, farmer/operator characteristics, specific topography, soil productivity, and soil texture, and weather. Furthermore, because of the nature of the relationships hypothesized, it is important to realize that they are not meant to imply causality. Rather, they define identities that describe the relationships of conservation tillage adoption and the extent of conservation tillage adoption to the inputs used in corn production, farmer/operator characteristics, topography, soil productivity, etc.

Finally, before presenting the estimation results, a few additional comments on the data and estimation are needed. First, the estimation of the Cragg and Heckman models are carried via maximum likelihood with the algorithm defined by Davidson, Fletcher, and Powell (Davidson and MacKinnon [1993]). Next, the data that are farm size dependent (e.g., total fertilizer, pesticide, and fuel expenditures, total yield, quantity of water applied, etc.) are divided by the number of acres planted to mitigate any potential effects of heteroscedasticity on the estimates due to farm size. Third, since the temperature and rainfall for 1987 were not abnormal, use of the actual data as a proxy for expected temperature and precipitation[82] is acceptable.

[82] Since the farmer does not know before making a conservation tillage decision what the weather (temperature and rainfall) will be like, he or she must base the decision on expectations, to the extent weather is a factor in the decision. The actual approach to expectations is not central to the current analysis. To minimize obfuscation, a common, straightforward approach to expectations formation is employed (Intrilligator [1978]).

Table 10.3a contains the coefficient estimates together with standard errors for the independent Cragg model and the complete dominance model for the no till production system and Table 10.3b contains coefficient estimates and standard errors of the estimates for mulch tillage.[83] Also, as noted previously, because differences in the probability of selection introduces bias in the estimates of the coefficients and their variances obtained via conventional estimation techniques, a weighted procedure in which the weights are equal to the inverse of the probability of selection must be used. Both of the models fit the data reasonably well for both no till and mulch tillage. The no till adoption equation correctly classifies (as adopters or nonadopters) about 82 percent of the observations using the (0.5,0.5) criterion while the mulch tillage correctly classifies about 86 percent of the observations. For this criterion, a correct classification means that the predicted probability of conservation tillage adoption is equal to or greater than 0.5 for an actual adopter and below 0.5 for a nonadopter (Maddala [1983]).

Table 10.3a. Parameter estimates of the independent Cragg model and the complete dominance model for no till.
Standard errors of the estimates are in parentheses.

Variable	Cragg Model	Complete Dominance
(A) Conservation Tillage Adoption		
CONSTANT	-7.315* (2.023)	-7.854* (2.178)
FEXP	0.041* (0.017)	0.047* (0.021)
PEXP	0.029* (0.007)	0.022* (0.010)
FUEL	-0.012* (0.004)	-0.014* (0.006)
LABOR	0.011 (0.015)	0.004 (0.063)
OWNLABOR	-0.107 (0.301)	-0.085* (0.033)
CUSTOM	-2.811* (1.312)	-2.887* (0.706)
IRRIGATE	-0.001 (0.207)	-0.016 (0.037)
WATER	0.251 (0.319)	0.195 (0.402)
SEED	0.003 (0.061)	0.001 (0.317)

[83] Note that in order to avoid problems in interpreting the results, farms that adopted mulch tillage exclusively are omitted from the sample used in the estimation of the no till models and those that adopted no till exclusively are omitted from the sample used in the estimation of the mulch tillage models. The impact of the 33 farms that adopted both no till and mulch tillage on some portion of their cropland is captured by introducing a proportion of cropland devoted to a specific conservation tillage practice variable in the extent of conservation tillage adoption relationship.

Table 10.3a. Parameter estimates of the independent Cragg model and the complete dominance model for no till. (continued)

NOPEST	0.415 (0.312)	0.333 (1.760)
YIELD	0.003 (0.021)	0.001 (0.022)
ARP	0.396 (0.760)	0.489 (0.731)
TOTALACRES	0.001 (0.007)	0.001 (0.074)
FARMTYPE	0.461* (0.126)	0.378* (0.134)
OWNTYPE1	0.251 (0.727)	0.171 (0.276)
OWNTYPE2	0.413 (0.460)	0.311 (0.581)
EDUCATION1	-0.944 (1.238)	-0.726 (1.184)
EDUCATION2	-0.611 (0.653)	-0.474 (0.727)
EDUCATION3	-0.369 (0.453)	-0.279 (0.391)
AGE1	0.116 (0.760)	0.002 (0.210)
AGE2	0.073 (0.321)	0.063 (0.251)
AGE3	0.054 (0.195)	-0.044 (0.189)
AGE4	-0.041 (0.272)	0.057 (0.231)
SLOPE	0.092* (0.026)	0.079* (0.027)
TEXTURE1	-0.078 (0.298)	-0.044 (0.315)
TEXTURE2	-0.233 (0.275)	-0.189 (0.271)
SOILPROD1	-0.333* (0.117)	-0.234* (0.113)
SOILPROD2	0.171* (0.053)	0.132* (0.047)
AVGTEMP	0.055 (0.057)	0.044 (0.037)
AVGRAIN	0.153* (0.057)	0.127* (0.050)
(B) Extent of Conservation Tillage Adoption		
CONSTANT	-0.222 (0.137)	-0.376 (0.599)
FEXP	0.078* (0.027)	0.071* (0.022)
PEXP	0.059* (0.012)	0.062* (0.025)
FUEL	-0.009* (0.003)	-0.015* (0.006)
LABOR	-0.004 (0.004)	-0.009 (0.014)
OWNLABOR	-0.002 (0.014)	0.054 (0.107)
CUSTOM	0.143* (0.062)	0.293* (0.053)
IRRIGATE	-0.005 (0.008)	-0.013 (0.046)
WATER	0.099 (0.376)	0.073 (0.176)
SEED	-0.006 (0.072)	0.011 (0.271)
NOPEST	0.112 (0.342)	0.027 (0.078)
YIELD	0.001 (0.051)	0.005 (0.017)
ARP	0.039 (0.046)	0.119 (0.118)
TOTALACRES	-0.001 (0.014)	0.001 (0.052)
FARMTYPE	0.016* (0.007)	0.009* (0.002)
OWNTYPE1	-0.018 (0.017)	-0.035 (0.059)
OWNTYPE2	-0.027 (0.029)	-0.161 (0.310)
EDUCATION1	0.017 (0.089)	0.044 (0.233)

Table 10.3a. Parameter estimates of the independent Cragg model and the complete dominance model for no till. (continued)

EDUCATION2	0.018 (0.051)	0.072 (0.153)
EDUCATION3	0.012 (0.028)	0.076 (0.088)
AGE1	-0.087 (0.173)	-0.037 (0.085)
AGE2	-0.003 (0.014)	-0.014 (0.044)
AGE3	0.009 (0.116)	-0.041 (0.075)
AGE4	0.030 (0.177)	-0.026 (0.058)
SLOPE	0.043* (0.020)	0.077* (0.025)
TEXTURE1	-0.007 (0.029)	0.049 (0.065)
TEXTURE2	-0.005 (0.017)	-0.018 (0.069)
SOILPROD1	0.002 (0.090)	-0.007 (0.015)
SOILPROD2	-0.021 (0.039)	-0.086 (0.093)
AVGTEMP	0.002 (0.023)	0.004* (0.002)
AVGRAIN	0.009* (0.003)	0.005* (0.002)
PMT	-0.051* (0.017)	-0.197* (0.081)
Sigma	1.717 (0.314)	
Log likelihood	252.2	(A)+(B)=267.6

* Significant at the 5 percent level or better.

Table 10.3b. Parameter estimates of the independent Cragg model and the complete dominance model for mulch tillage. Standard errors of the estimates are in parentheses.

Variable	Cragg Model	Complete Dominance
(A) Conservation Tillage Adoption		
CONSTANT	-5.307* (2.011)	-0.021 (2.064)
FEXP	0.018 (0.037)	0.001 (0.008)
PEXP	0.022* (0.009)	0.045* (0.015)
FUEL	-0.018* (0.007)	-0.009* (0.002)
LABOR	0.008 (0.053)	0.007 (0.057)
OWNLABOR	-0.082 (0.092)	-0.054* (0.020)
CUSTOM	-2.744* (1.144)	-2.554* (0.870)
IRRIGATE	-0.014 (0.038)	0.001 (0.001)
WATER	0.316 (0.520)	0.207 (0.496)
SEED	0.002 (0.073)	0.001 (0.019)
NOPEST	0.319 (0.427)	-0.610 (1.341)
YIELD	0.002 (0.025)	0.001 (0.009)
ARP	0.399 (0.689)	0.611 (0.691)
TOTALACRES	0.003 (0.042)	0.001 (0.743)

Table 10.3b. Parameter estimates of the independent Cragg model and the complete dominance model for mulch tillage.

FARMTYPE	0.357* (0.117)	0.248* (0.101)
OWNTYPE1	0.195 (0.633)	-0.070 (0.236)
OWNTYPE2	0.326 (0.517)	-0.336 (0.205)
EDUCATION1	-0.732 (1.064)	1.243 (0.980)
EDUCATION2	-0.476 (0.756)	1.140 (0.769)
EDUCATION3	-0.287 (0.538)	0.518 (0.407)
AGE1	0.097 (0.326)	0.013 (0.279)
AGE2	0.065 (0.167)	0.212 (0.297)
AGE3	-0.039 (0.188)	-0.165 (0.180)
AGE4	0.023 (0.208)	0.075 (0.199)
SLOPE	0.076* (0.025)	0.138* (0.029)
TEXTURE1	-0.060 (0.281)	0.105 (0.286)
TEXTURE2	-0.189 (0.250)	0.338 (0.200)
SOILPROD1	-0.255* (0.128)	-0.240* (0.119)
SOILPROD2	0.134* (0.046)	0.056* (0.024)
AVGTEMP	0.041 (0.030)	0.029 (0.021)
AVGRAIN	0.115* (0.051)	0.207* (0.055)
(B) Extent of Conservation Tillage Adoption		
CONSTANT	0.015 (1.895)	-1.008 (0.798)
FEXP	0.002 (0.072)	0.036 (0.027)
PEXP	0.038* (0.016)	0.033* (0.012)
FUEL	-0.007* (0.002)	-0.042* (0.015)
LABOR	-0.006 (0.055)	0.003 (0.017)
OWNLABOR	-0.006 (0.019)	0.012 (0.007)
CUSTOM	-6.255* (2.592)	-2.793* (0.873)
IRRIGATE	-0.013 (0.027)	-0.016 (0.115)
WATER	0.165 (0.388)	0.111 (0.195)
SEED	0.001 (0.106)	-0.015 (0.340)
NOPEST	-0.594 (1.370)	0.809 (0.502)
YIELD	0.002 (0.003)	0.008 (0.006)
ARP	0.566 (0.648)	-0.009 (0.233)
TOTALACRES	0.001 (0.065)	0.002 (0.003)
FARMTYPE	0.243* (0.103)	0.139* (0.039)
OWNTYPE1	-0.072 (0.215)	0.025 (0.063)
OWNTYPE2	-0.303 (0.263)	-0.015 (0.078)
EDUCATION1	1.177 (1.766)	0.311 (0.409)
EDUCATION2	1.066 (0.756)	0.259 (0.263)
EDUCATION3	0.495 (0.387)	0.127 (0.138)
AGE1	0.078 (0.200)	0.018 (0.094)
AGE2	0.016 (0.254)	0.032 (0.051)
AGE3	-0.180 (0.293)	0.109 (0.187)

Table 10.3b. Parameter estimates of the independent Cragg model and the complete dominance model for mulch tillage. (continued)

AGE4	0.065 (0.192)	-0.047 (0.061)
SLOPE	0.136* (0.028)	0.147* (0.060)
TEXTURE1	0.101 (0.257)	-0.073 (0.086)
TEXTURE2	0.316 (0.294)	0.100 (0.063)
SOILPROD1	-0.228* (0.109)	-0.147* (0.051)
SOILPROD2	0.502* (0.213)	0.411* (0.120)
AVGTEMP	0.018 (0.022)	0.023 (0.019)
AVGRAIN	0.193* (0.054)	0.036* (0.009)
PNT	-0.083* (0.004)	-0.073* (0.015)
Sigma	1.440 (0.281)	
Log likelihood	117.3	(A)+(B)=109.5

Significant at the 5 percent level or better.

A maximum likelihood ratio test accepts the hypothesis that the independent Cragg model is an acceptable alternative to the Cragg model with dependence for both no till ($\chi2 = 1.03$) and mulch tillage ($\chi2 = 2.35$).[84] This indicates that the conservation tillage adoption decision and the extent of conservation tillage adoption decision are not made simultaneously. A likelihood ratio test also indicates that the complete dominance model is an acceptable form of the Heckman model for both no till ($\chi2 = 1.73$) and mulch tillage ($\chi2 = 0.81$).[85] Since this likelihood ratio test is also a test for sample selection bias, the latter is not a significant problem for either no till or mulch tillage production systems (Dhrymes [1986]).

Based on the results of the likelihood ratio tests, it is concluded that the independent Cragg and the complete dominance models are acceptable formulations for modeling the factors that are interrelated with the conservation tillage adoption decisions in the sample. Both of these models imply the conservation tillage adoption and the extent of adoption decisions are not made simultaneously. The major difference lies in the treatment of zero observations. The Cragg model implies that the conservation tillage adoption decision should be estimated over the entire population including both conservation

[84] Both of the χ^2 tests are performed with one degree of freedom. The critical value at the 5 percent level is 3.84.

[85] Again both of the χ^2 tests are performed with one degree of freedom.

tillage adopters and nonadopters and the complete dominance model implies the relevant population consists only of conservation tillage adopters. For the no till and the mulch tillage systems, there are a comparable number of statistically significant coefficients (at the 5 percent level) for both the Cragg and complete dominance models for the conservation tillage adoption and the extent of adoption equations. Moreover, coefficient estimates, especially for those that are statistically significant, have similar signs across the two models. Note, however, making a direct comparison concerning such things as the order of magnitude of the coefficient estimates is inappropriate because the complete dominance model only includes conservation tillage adopters and the Cragg model uses all observations in the sample. Because a clear-cut decision cannot be made as to which model is preferable, the subjective preference is to choose the most general of the models - the Cragg model. The Cragg model may be preferable because, unlike the complete dominance model, it assumes that a current conservation tillage nonadopter could be induced to adopt one or more of the conservation tillage practices if something changed.

What do the results say about the relationship between conservation tillage adopters and the various FCRS variables including expenditures on and use of factor inputs, farmer/operator characteristics, specific topography, soil productivity, and soil texture, and weather? Consider the empirical results for the two conservation tillage systems in turn.

(A) No Till Production System Results

A farm that is a cash grain enterprise is about 24 percentage points more likely to adopt no till than, say, a dairy farm while the type of ownership (e.g., single owner versus a partnership) of the farm has no effect on the conservation tillage adoption decision.

The slope of the cropland is an important (statistically significant) factor associated with the adoption of no till. As the slope of the cropland increases, the propensity for water runoff and soil erosion increases[86] although the precise functional relationship is subject to debate (Uri and Hyberg [1990]). As noted previously, the no till production practice can be significant in reducing

[85] Again both of the χ^2 tests are performed with one degree of freedom.

[86] This is reflected by the general use of the Universal Soil Loss Equation (USLE) which contains a rainfall erosion index. The USLE was designed to capture the relationship between stream sediment loading and storm intensity (Wischmeier and Smith [1978]). The Appendix to Chapter 7 discusses this equation.

water runoff and soil erosion from fields. Farmers in the sample, on average, appear to be using this information. The estimation results suggest that a one percent increase in the slope of the cropland is associated with about a 44 percentage point increase in the likelihood that no till will be adopted.[87]

Average rainfall but not average temperature is associated with a small, but statistically significant (at the 5 percent level) greater likelihood that no till is adopted. Thus, a 10 percent greater level of rainfall is associated with a 4 percentage point greater likelihood that no till is adopted on some portion of the cropland. This is additional confirmation that a farmer adopting no till is trying to mitigate water runoff and soil erosion since rainfall and soil erosion and, hence, stream sediment loading are inexorably intertwined (Novotny and Chester [1981]).

A farmer who adopts one conservation tillage practice is less likely to adopt another conservation tillage practice. That is, the farmer who adopts no till is 15 percentage points less likely to adopt mulch tillage. This result seems an anomaly since adoption of no till on a portion of the cropland does not inherently preclude adoption of mulch tillage on another portion of the cropland. The result and its attendant implications is an issue that needs to be explored further.

Finally, no till adopters spend more on fertilizer and pesticides but less on fuel than nonadopters. Conservation tillage changes soil properties in ways that affect plant growth (Phillips et al. [1980]). Plant nutrients in no till soils are more stratified than those in soils under reduced or conventional tillage. Nutrients tend to concentrate in the upper portion of the soil profile. Additionally, soil under no till practices, which leaves a surface mulch, is often 3 to 4 degrees Celsius cooler in late spring than soil under conventional tillage. In the spring, the cooler temperatures can slow early season plant growth (National Research Council [1989]). To compensate for these effects of no till, additional fertilizer is required. Recognition of this fact by the farmer is reflected in the estimation results. On average, a farmer who adopts no till is 23 percentage points more likely to have higher fertilizer costs than a nonadopter as indicated by the additional expenditures. Additionally, for a no till adopter who applies additional fertilizer, a one percent increase in acreage under con-

[87] Note this value was computed based on the mean of the slope variable over the entire sample. Other response values are similarly computed.

servation tillage is associated with a 1.3 percent greater expenditure on fertilizer.

With regard to pesticides, the reduced mechanical control of perennial weeds through conventional tillage necessitates an increased reliance on herbicides. Additionally, crop residue on the top of soil resulting from no till provides a favorable habitat to some pests and some plant diseases overwinter in crop residues left on the soil and above and below ground insects survive (Hendrix et al. [1986]). Thus, consistent with a priori expectations, a farmer adopting no till is more likely to incur increased expenditures concurrent with an increased use of pesticides. A farmer who adopts no till is 57 percentage points more likely to have a higher expenditure on pesticides than a nonadopter. Moreover, under a no till system, per acre expenditure on pesticides (evaluated at the mean of the sample) is 9.5 percent greater than for a conventional tillage system, all other things given (such as expenditures on other factor inputs, soil productivity, slope of the cropland, etc.).

Coincident with the greater likelihood of increased expenditures on pesticides for no till adopters is the reduced likelihood that adopters will contract for custom applications of pesticides. That is, the more extensive use of pesticides associated with no till results in a lower requirement for custom applications. A farmer who adopts no till is 25 percentage points more likely to have reduced expenditures for custom pesticide applications. Additionally, for a no till adopter who uses custom pesticide applications less, a one percent increase in acreage under conservation tillage is associated with a 2.7 percent lower expenditure on custom pesticide applications.

Fuel expenditures are expected to be less under no till than they are under conventional tillage because there will be a significant reduction in the use of diesel fuel and gasoline for tillage even though this is partially offset by the relatively small increase in fuel use for the additional fertilizer and pesticide applications noted previously (Uri and Day [1992]). In a relative sense, fuel used for tillage accounts for approximately 11 percent of total direct farm energy use while fertilizer application accounts for 0.7 percent and pesticide application accounts for 0.8 percent (Economic Research Service [1987]). Thus, a reduction in farm fuel use due to the reduction in or elimination of tillage would be expected to be substantially in excess of the increase in fuel use for fertilizer and pesticide applications. The data show that, consistent with this expectation, a farmer who adopts no till is 16 percentage points more likely to have lower fuel expenditures than a nonadopter. Moreover, for a no till adopter who uses less fuel, a one percent increase in acreage under conservation tillage is associated with a 2.8 percent lower expenditure on fuel.

A number of factors including expenditures on some inputs and farm and farm owner/operator characteristics are found not to be associated with the adoption or nonadoption of the no till production practice. For example, the age and education level of the farmer/operator is not statistically significantly associated with the adoption of no till. Thus, the suggestion that younger, better educated farmers are more cognizant of the off-farm effects of agricultural production and hence are more inclined to adopt a conservation tillage practice is not supported by the empirical results (Ferguson and Yee [1995] and Ogg [1992]).

The productivity of the soil as measured by average yield across farms in a county does not appear to have any identifiable impact on the decision to adopt no till. While clearly there is a positive relationship between highly erodible, medium to high productivity soils and the no till adoption decision (Heimlich and Bills [1984]), on average and across all soils, soil productivity is not correlated with the no till adoption decision.

The texture of the soil, the total acres planted, the number of acres in the acreage reduction program, the extent of irrigation, and the proportion of acres not receiving any pesticide treatment are not interrelated with the adoption of no till. That is, such things as the size of the farm, the extent of government farm program participation, and the type of soil appear, on average, to be unimportant in the conservation tillage adoption decision. This is surprising since the adoption of alternative farming systems has been shown to be a intertwined with government farm program participation and farm size (Dobbs et al. [1988] and Ogg [1990]). This issue is clearly deserving of further study.

Hired labor and owner/operator labor are two additional factors that are not significantly correlated with the conservation tillage adoption decision. That is, the decision to adopt no till farming has no relationship to the use of hired labor or to the amount of labor expended by the farmer/operator. Also, no till farming is neither more nor less labor intensive than farming relying on conventional tillage practices (Office of Technology Assessment [1986]). Thus, it is not possible to conclude that no till is a labor saving production practice from these data.

(B) Mulch Tillage Production System Results

Table 10.3b presents information about the relationship between mulch tillage adopters and the various FCRS variables including expenditures on and use of factor inputs, farmer/operator characteristics, specific topography, soil productivity, and soil texture, and weather. The pattern of statistical significance of coefficient estimates of the variables is very similar to that observed for the no till production practice and the reasons are analogous. Consequently, an extensive discussion of the results is not required. Just the highlights will be noted. A farm that is a cash grain enterprise is approximately 18 percentage points more likely to adopt mulch tillage than, say, a beef or hog farm while the type of ownership of the farm has no effect on the conservation tillage adoption decision. A farm with high productivity soil is about 8 percentage points less likely to adopt mulch tillage relative to a farm with average productivity soil while a farm with low productivity is about 8 percentage points more likely to adopt mulch tillage.

The slope of the cropland is a statistically significant (at the 5 percent level) factor associated with the adoption of mulch tillage. The results suggest that a 1 percent increase in the slope of the cropland is associated with about a 25 percentage point increase in the likelihood that mulch tillage will be adopted.

Average rainfall but not average temperature is associated with a statistically significant (at the 5 percent level) greater likelihood that mulch tillage is adopted. A 10 percent greater level of rainfall is associated with a 42 percentage point greater likelihood that mulch tillage is adopted on some portion of the cropland. This effect is substantially greater than it is for the no till production practice results reported previously. The precise reason for the order of magnitude difference is properly a subject for future study.

A farmer who adopts mulch tillage is 12 percentage points less likely to adopt no till. This inverse relationship is the same as that observed for no till. Clearly, the adoption of no till on a portion of cropland means that less cropland is available for mulch tillage but one should not necessarily witness a fall in the probability of the adoption of mulch tillage because of this. Why the adoption of one conservation tillage practice impacts the likelihood that another one will be adopted is an anomaly that is in need of further exploration.

Finally, mulch tillage adopters spend more on pesticides but less on fuel than nonadopters. Consistent with expectations, a farmer adopting mulch tillage is more likely to incur increased expenditures coincident with an increased use of pesticides. A farmer who adopts mulch tillage is 31 percentage points more likely to have a higher expenditure on pesticides than a nonadopter. Additionally, under a mulch tillage, per acre expenditure on pesticides (evaluated at the mean of the sample) is 12.9 percent greater than for a conventional tillage system.

Along with the greater likelihood of increased expenditures on pesticides for mulch tillage is the reduced likelihood that adopters will contract for custom applications of pesticides. As was the case for no till, the more extensive use of pesticides associated with mulch tillage results in a lower requirement for custom applications. A farmer who adopts mulch tillage is 16 percentage points more likely to have reduced expenditures for custom pesticide applications. Additionally, for a mulch tillage adopter who uses custom pesticide applications less, a 1 percent increase in acreage under conservation tillage is associated with a 0.99 percent lower expenditure on custom pesticide applications.

Consistent with expectations, a farmer who adopts mulch tillage is 12 percentage points more likely to have lower fuel expenditures than a nonadopter. Moreover, for a mulch tillage adopter who uses less fuel, a one percent increase in acreage under conservation tillage is associated with a 1.1 percent lower expenditure on fuel.

One difference between the estimation results for no till and mulch tillage is that while there is an increased likelihood of greater expenditure on fertilizer associated with the adoption of no till, such is not the case with mulch tillage. There is no statistically significant relationship between the adoption of mulch tillage versus its nonadoption and the likelihood of increased fertilizer expenditure. This is probably reflecting the fact that under a mulch tillage production practice, the soil is disturbed thereby serving to redistribute the nutrients that tend to concentrate in the upper portion of the soil profile (Bull [1993]).

Several factors including expenditures on some inputs and farm and farm owner/operator characteristics are found not to be associated with the adoption or nonadoption of the mulch tillage production practice. Thus, the age and education level of the farmer/operator is not statistically significantly associated with the adoption of mulch tillage. Consequently, as was the case for the no till production system, the suggestion that younger, better educated farmers are more aware of the environmental impacts of agricultural production than are their older, less well-educated counterparts and hence are more likely to adopt a conservation tillage practice is not confirmed by the empirical results. The texture of the soil, the productivity of the soil, the total acres planted, the number of acres in the acreage reduction program, the extent of irrigation, and the proportion of acres not receiving any pesticide treatment are not interrelated with the adoption of mulch tillage. That is, such things as the size of the farm, the extent of government farm program participation, and the type of soil are unimportant in the conservation tillage adoption decision.

Expenditures on hired labor and the amount of owner/operator labor expended are additional factors not significantly interrelated with the conservation tillage adoption decision. That is, the decision to adopt mulch tillage farming has no relationship to expenditures on hired labor or the quantity labor expended by the farmer/operator. Mulch tillage farming is neither more nor less labor intensive than farming relying on conventional tillage practices.

CONCLUSION

Conservation tillage practices do have beneficial environmental effects in terms of reduced water runoff and mitigated soil erosion. There continues to be a question, however, as to their overall effectiveness in reducing the impact of agricultural production on the environment especially with regard to pesticides and fertilizer. While it is generally recognized that pesticide use will increase under conservation tillage, what has not previously been adequately studied is the extent to which it will increase. Additionally, the impact of conservation tillage on fertilizer use is uncertain.

To investigate these issues, it is assumed that the conservation tillage adoption decision is a two step procedure - the first is the decision whether or not to adopt a conservation tillage production system and second is the decision on the extent to which conservation tillage should be used. With this formulation of the problem, a double hurdle modeling approach is appropriate. By using this approach, it is possible to test whether decisions about conservation tillage adoption and the extent of adoption are separate, endogenous choices. Also, if it is credible that the conservation tillage adoption and extent of adoption decisions are made simultaneously then the equation error terms are correlated. Cragg and Heckman (or dominance) models were estimated with and without the assumption of dependent error terms. The Cragg model with an independent error term was the preferred model. Unlike the dominance models, the Cragg assumes all zero observations represent standard corner solutions.

Based on farm-level data on corn production in the United States for 1987, the profile of a farm on which conservation tillage was adopted is that the cropland had above average slope and experienced above average rainfall, the farm was a cash grain enterprise, and it had an above average expenditure on pesticides and a below average expenditure on fuel and a below average expenditure on custom pesticide application. Additionally, for a farm adopting a no till production practice, an above average expenditure was made on fertilizer.

These results translate into a number of issues concerning expenditures on pesticides and fertilizer with their attendant environmental implications. First, with the adoption of conservation tillage, per acre expenditure on pesticides (evaluated at the mean of the sample) under no till is 9.5 percent greater than for a conventional tillage system and 12.9 percent greater under mulch tillage, all other things given (including farm type, soil productivity, rainfall, labor inputs, etc.).

The increase in expenditures on pesticides has a number of implications. One of these has to do with the adoption of conservation tillage on highly erodible land[88] not currently subject to conservation compliance. Such adoption has been suggested as a policy option to reduce soil erosion (National Research Council [1989] and Zinn [1993]). In 1994, 98 million of the 149 million acres of cropland in the United States classified as highly erodible land had fully implemented conservation compliance systems. Thus, there remains 51 million acres on which one or more of the conservation tillage practices can be adopted (Economic Research Service [1997]). This represents approximately 18 percent of the total acreage planted to major field crops in 1994 (Economic Research Service [1995]). If it is assumed that all crops planted on the acreage not currently in compliance must use either a no till or mulch tillage system or combination of the two and if it is further assumed that the empirical results obtained for corn production are applicable to all major field crops, then pesticide expenditures can be expected to rise. Based

[88] Determinations made by the Natural Resources Conservation Service of the U.S. Department of Agriculture staff include cropland in fields that have at least one-third or 50 acres (whichever is less) of highly erodible soils. These are soils with a natural erosion potential at 8 times their tolerance value. The tolerance value is the rate of soil erosion above which long-run soil productivity may be depleted (Economic Research Service [1986]).

on the means values in Table 10.2, pesticide expenditures will rise by between \$0.45 and \$0.61 per acre while custom pesticide application expenditures will decrease by between \$0.01 and \$0.03 per acre as conservation tillage is adopted on the remaining highly erodible land. Aggregate pesticide expenditures will rise by between \$22 and \$30 million dollars per year. With the average price of pesticides at \$4.45 per pound of active ingredient in 1994 (Economic Research Service [1995]), total pesticide use will increase by between 4.9 and 6.7 million pounds of active ingredients.

Finally, if no till is adopted exclusively on all highly erodible land not currently in compliance, then there will be a rise in per acre pesticide expenditures of about \$0.45, a reduction in per acre custom pesticide applications of \$0.03, and a per acre rise in fertilizer expenditures of \$0.18. In this instance, if a no till system is adopted on all remaining highly erodible land, total fertilizer expenditures will increase by \$9.2 million. With the price of fertilizer in 1994 at \$0.12 per pound (Economic Research Service [1995]), this means that an additional 77 million pounds of fertilizer would be applied.

REFERENCES

Anderson, J., J. Dillon, and B. Hardarker, *Agricultural Decision Analysis*, The Iowa State University Press, Ames, Iowa, 1977.

Baker, J.L., "Agricultural Areas as Nonpoint Sources of Pollution," in M. Overcash and J. Davidson (eds.), *Environmental Impact of Nonpoint Source Pollution*, Ann Arbor Science Publications, Ann Arbor, MI, 1980, pp. 275-310.

Baker, J.L., "Hydrologic Effects of Conservation Tillage and Their Importance to Water Quality," in T. Logan, J. Davidson, J. Baker, and M. Overcash (eds.), *Effects of Conservation Tillage on Groundwater Quality: Nitrates and Pesticides*, Lewis Publishers, Chelsea, MI, 1987, pp. 113-124.

Baker, J., and H. Johnson, "The Effect of Tillage Systems on Pesticides in Runoff from Small Watersheds," *Transactions of the American Society of Agricultural Engineers*, Vol. 22 (1979), pp. 554-559.

Baumol, W.J., Economic Theory and Operations Analysis, Prentice-Hall, Inc., Englewood Cliffs, NJ 1972.

Blundell, R., and C. Meghir, "Bivariate Alternative to the Univariate Tobit Model," *Journal of Econometrics*, Vol. 33 (1987), pp. 179-200.

Bruce, R., G. Langdale, L. West, and W. Miller, "Surface Soil Degradation and Soil Productivity Restoration and Maintenance," *Soil Science Society of America Journal*, Vol. 59 (1995), pp. 654-660.

Bull, L., Residue and Tillage Systems for Field Crops, *SR AEGS 9310*, Resources and Technology Division, Economic Research Service, U.S. Department of Agriculture, Washington, July 1993.

Chamberlain, G., "Panel Data," in *Handbook of Econometrics*, Z. Griliches and M. Intriligator (eds.), North-Holland Publishing Company, Amsterdam, 1986.

Cragg, J., "Some Statistical Models for Limited Dependent Variables with Applications to the Demand for Durable Goods," *Econometrica*, Vol. 39 (1971), pp. 829-844.

Davidson, R., and J. MacKinnon, *Estimation and Inference in Econometrics*, Oxford University Press, Oxford, 1993.

Deaton, A., and M. Irish, "Statistical Models for Zero Expenditures in Household Budgets," *Journal of Public Economics*, Vol. 23 (1986), pp. 59-80.

Dhrymes, P., "Limited Dependent Variables," in *Handbook of Econometrics*, Z. Griliches and M. Intrilligator (eds.), North Holland Publishing Company, 1986.

Dick, W.A., and T.C. Daniel, "Soil Chemical and Biological Properties as Affected by Conservation Tillage: Environmental Implications," in T. Logan, J. Davidson, J. Baker, and M. Overcash (eds.), *Effects of Conservation Tillage on Groundwater Quality: Nitrates and Pesticides*, Lewis Publishers, Chelsea, MI, 1987, pp. 315-339.

Diebel, P., D. Taylor, S. Batie and C. Heatwole, "Low Input Agriculture as a Groundwater Protection Strategy," *Water Resources Bulletin*, Vol. 28 (1992), pp. 755-761.

Diebel, P., D. Taylor, and S. Batie, "Barriers to Low-Input Agriculture Adoption," *American Journal of Alternative Agriculture*, Vol. 8 (1993), pp. 120-127.

Dobbs, T., M. Leddy, and J. Smolik, "Factors Influencing the Economic Potential for Alternative Farming Systems," *American Journal of Alternative Agriculture*, Vol. 8 (1988), pp. 26-34.

Economic Research Service, An Economic Analysis of USDA Erosion Control Programs: A New Perspective, *AER 560,* U.S. Department of Agriculture, Washington, August 1986.

Economic Research Service, Energy and U.S. Agriculture: State and National Fuel Use Tables, *AGES861121,* U.S. Department of Agriculture, Washington, March 1987.

Economic Research Service, Agricultural Resources and Environmental Indicators, *AH 705,* U.S. Department of Agriculture, Washington, December 1994.

Economic Research Service, Agricultural Outlook, *AO 219,* U.S. Department of Agriculture, Washington, June 1995.

Edwards, W., M. Shipitalo, L. Owens, and W. Dick, "Factors Affecting Preferential Flow of Water and Atrazine Through Earthworm Burrows under Continuous No-Till Corn," *Journal of Environmental Quality,* Vol. 22 (1993), pp. 225-241.

Fawcett, R.S., "Overview of Pest Management for Conservation Tillage Systems," in T. Logan, J. Davidson, J. Baker, and M. Overcash (eds.), *Effects of Conservation Tillage on Groundwater Quality: Nitrates and Pesticides,* Lewis Publishers, Chelsea, MI, 1987, pp. 6-19.

Fawcett, R., D. Tierney, and B. Christensen, "Impact of Conservation Tillage on Reducing Runoff of Pesticides into Surface Water," *Journal of Soil and Water Conservation,* Vol. 49 (1994), pp. 49-56.

Ferguson, W., and J. Yee, "A Logit Model of Cotton Producer Participation in Professional Scout Programs," *Journal of Sustainable Agriculture,* Vol. 5 (1995), pp. 87-96.

Fuller, W., "Least Squares and Related Analyses for Complex Survey Design," *Survey Methodology,* Vol. 10 (1984), pp. 97-118.

Fuller, W., and M.A. Hidiroglou, "Regression Estimation After Correcting for Attenuation," *Journal of the American Statistical Association,* Vol. 73 (1978), pp. 99-104.

Glenn, S., and J. Angle, "Atrazine and Simazine in Runoff from Conventional and No-Till Corn Watersheds," *Agricultural Ecosystems and Environment,* Vol. 18 (1987), pp. 221-244.

Hall, J., N. Hartwig, and L. Hoffman, "Cyanzine Losses in Runoff from No-Tillage Corn in "Living" and Dead Mulches Versus Unmulched, Conventional Tillage," *Journal of Environmental Quality,* Vol. 13 (1984), pp. 105-110.

Heckman, J.J., "Sample Selection Bias as a Specification Error," *Econometrica*, Vol. 47 (1979), pp. 153-161.

Hendrix, P., R. Parmelee, D. Crossley, D. Coleman, E. Odum, and P. Groffman, "Detritus Food Webs in Conventional and No Till Agroecosystems," *Bioscience*, Vol. 36 (1986), pp. 374-380.

Holland, E.A., and D.C. Coleman, "Litter Placement Effects on Microbial and Organic Matter Dynamics in an Agroecosystem," *Ecology*, Vol. 68 (1987), pp. 425-433.

Intrilligator, M., *Econometric Models, Techniques, and Applications*, Prentice-Hall, Inc., Englewood Cliffs, NJ, 1978.

Ismail, I., R. Blevins, and W. Frye, "Long Term No-Tillage Effects on Soil Properties and Continuous Corn Yields," *Soil Science Society of America Journal*, Vol. 58 (1994), pp. 193-198.

Kellogg, R., M. Maizel, and D. Goss, *Agricultural Chemical Use and Groundwater Quality: Where Are the Potential Problems*, National Center for Resource Innovations, Washington, DC, 1994.

Keeney, R., and H. Raiffa, *Decisions with Multiple Objectives*, John Wiley and Sons, Inc., New York, 1976.

Lee, E.S., R.N. Forthofer, and R.J. Lorimor, *Analyzing Complex Survey Data*, Sage University Press, Newbury Park, CA, 1989.

Lee, L., and G.S. Maddala, "The Common Structure of Tests for Selectivity Bias, Serial Correlation, Heteroscedasticity, and Nonnormality in the Tobit Model," *International Economic Review*, Vol. 32 (1985), pp. 238-251.

Lipton, K., and S. Pollack, *A Glossary of Food and Agricultural Policy Terms*, Economic Research Service, U.S. Department of Agriculture, Washington, DC, November 1989.

Maddala, G.S., *Limited Dependent and Qualitative Variables in Econometrics*, Cambridge University Press, Cambridge, 1983.

National Research Council, *Alternative Agriculture*, National Academy Press, Washington, 1989.

National Research Council, *Soil and Water Quality: An Agenda for Agriculture*, National Academy Press, Washington, 1993.

Neumann, J. Von and O. Morgenstern, *Theory of Games and Economic Behavior*, 2nd ed., Princeton University Press, Princeton, NJ, 1947

Office of Technology Assessment, *Technology, Public Policy, and the Changing Structure of American Agriculture*, Congress of the United States, Washington, DC, March 1986.

Office of Technology Assessment, *Beneath the Bottom Line: Agricultural Approaches to Reduce Agrichemical Contamination of Groundwater*, Congress of the United States, Washington, DC, November 1990.

Ogg, C., "Farm Price Distortions, Chemical Use, and the Environment," *Journal of Soil and Water Conservation*, Vol. 45 (1990), pp. 45-48.

Ogg, C., "Addressing Information Needs to Support Sustainable Agriculture Policies," *Journal of Sustainable Agriculture*, Vol. 2 (1992), pp. 113-121.

Onstad, C., and W. Voorhees, "Hydrologic Soil Parameters Affected by Tillage," in T. Logan, J. Davidson, J. Baker, and M. Overcash (eds.), *Effects of Conservation Tillage on Groundwater Quality: Nitrates and Pesticides*, Lewis Publishers, Chelsea, MI, 1987, pp. 274-291.

Phillips, R., R. Blevins, G. Thomas, W. Frye, and S. Phillips, "No till Agriculture," Science, Vol. 208 (1980), pp. 1108-1113.

Pudney, S., *Modeling Individual Choice: The Econometrics of Corners, Kinks, and Holes*, Basil Blackwell, Ltd., London, 1989.

Reicosky, D., W. Kemper, G. Langdale, C. Douglas,, and P. Rasmussen, "Soil Organic Matter Changes Resulting from Tillage and Biomass Production," *Journal of Soil and Water Conservation*, Vol. 50 (1995), pp. 253-261.

Sander, K., W. Witt, and M. Barrett, "Movement of Triazine Herbicides in Conventional and Conservation Tillage Systems," in D.L. Weigmann (ed.), *Pesticides in Terrestrial and Aquatic Environments*, Virginia Water Resources Center, Virginia Polytechnic Institute and State University, Blacksburg, pp. 378-382.

Soil Conservation Service, *National Resource Inventory*, U.S. Department of Agriculture, Washington, 1982.

Stigler, G., *The Theory of Price*, Macmillan Publishing Company, New York, 1966.

Taylor, M., R. Adams, and S. Miller, "Farm Level Response to Agricultural Efficient Control Strategies: The Case of the Willamette Valley," *Agricultural and Economic Research*, Vol. 17 (1992), pp. 173-185.

Thorne, D.W., and M.D. Thorne, *Soil, Water and Crop Production*, AVI Publishing Company, Inc., Westport, Connecticut, 1979.

Uri, N.D., "Estimating the Agricultural Demand for Natural Gas and Liquefied Petroleum Gas in the Presence of Measurement Error in the Data," *International Journal of Energy Research*, Vol. 18 (1994), pp. 783-797.

Uri, N.D., and K. Day, "Energy Efficiency, Technological Change, and the Dieselization of American Agriculture in the United States," *Transportation Planning and Technology*, Vol. 16 (1992), pp. 221-231.

Uri, N.D., and B. Hyberg, "Stream Sediment Loading and Rainfall - A Look at the Issue," *Water, Air, and Soil Pollution*, Vol. 51 (1990), pp. 95-104.

Wagnet, R.J., "Processes Influencing Pesticide Loss with Water under Conservation Tillage," in T. Logan, J. Davidson, J. Baker, and M. Overcash (eds.), *Effects of Conservation Tillage on Groundwater Quality: Nitrates and Pesticides*, Lewis Publishers, Chelsea, MI, 1987, pp. 189-200.

Wauchope, R.D., "Effects of Conservation Tillage on Pesticide Loss with Water," in T. Logan, J. Davidson, J. Baker, and M. Overcash (eds.), *Effects of Conservation Tillage on Groundwater Quality: Nitrates and Pesticides*, Lewis Publishers, Chelsea, MI, 1987, pp. 201-215.

Webb, S., C. Ogg, W. Huang, Idling Erodible Cropland: Impacts on Production, Prices, and Government Costs, *AER 550*, Economic Research Service, U.S. Department of Agriculture, Washington, 1986.

Wischmeier, W., and D. Smith, Predicting Rainfall Erosion Losses - A Guide to Conservation Planning, *Agriculture Handbook No. 537*, U.S. Department of Agriculture, Washington, 1978.

Wood, C.W., J.H. Edwards, C.G. Cummins, "Tillage and Crop Rotation Effects on Soil Organic Matter in a Typic Hapludult of Northern Alabama," *Journal of Sustainable Agriculture*, Vol. 2 (1991), pp. 31-41.

Zilberman, D., and M. Marra, "Agricultural Externalities," *in Agricultural and Environmental Resources Economics*, G. Carlson, D. Zilberman, and J. Miranowski (eds.), Oxford University Press, Oxford, 1993.

Zinn, J., *Conservation Compliance: Status and Issues*, Congressional Research Service, Washington, 1993.

INDEX

A

agrichemicals · 1
agricultural policy · ix
agricultural pollutants · 3, 159, 185, 196
agricultural production · ix, 2, 3, 6, 7, 12, 35, 49, 51, 54, 67, 77, 78, 85, 92, 95, 96, 101, 108, 111, 118, 159, 181, 185, 195, 197, 198, 206, 217, 231, 232, 250, 257, 261, 280, 283
agricultural resource management · 1
Agroecosystem · 61, 154, 288
air quality · ix, 160, 195, 198
Alabama · 64, 89, 291
Alaska · 89
alfalfa · 47, 48, 50, 51, 76, 142
algae · 186, 196
Alternative · 5, 35, 44, 45, 60, 61, 62, 63, 122, 123, 124, 126, 130, 155, 193, 198, 223, 225, 251, 252, 254, 255, 286, 287, 289
alternative production practices · 3, 4, 17, 45, 46, 50, 86, 88, 96, 98, 99, 101, 102, 107, 109, 110, 145, 197
apples · 32, 79, 142
Atrazine · 34, 60, 81, 82, 191, 252, 287, 288
autoregressions · 244

B

base acreage · 110, 111, 112, 149
Beneficial Organisms · 136, 155
Biological Nitrogen Fixation · 62
biopesticides · 4, 133, 139, 140, 141, 142, 143, 146, 147, 148, 149, 152
Biopesticides · 5, 133, 143, 148, 152, 154
biotechnology · 139

C

caterpillars · 138
CCC · 110, 111
chemical use · ix, 2, 3, 4, 8, 9, 10, 11, 12, 42, 44, 54, 57, 73, 78, 83, 84, 92, 97, 107, 108, 110, 111, 112, 113, 114, 116, 117, 118, 119, 144, 146, 149, 257
Chemical Use · 5, 8, 57, 62, 67, 117, 125, 128, 136, 138, 288, 289
chemically-intensive crops · 113, 150
Choice of Production · 5, 67
chronic poisonings · 8
citrus · 31, 41, 138, 153
Clean Air Act · 71, 72, 144
cointegration · 6, 237, 239, 240, 243, 244, 245, 246, 250
commodities · 12, 24, 34, 70, 116, 118, 163, 189, 190, 201, 206, 240, 259, 265
compaction · ix, 53, 195, 202
conservation practices · 5, 6, 161, 162, 164, 172, 176, 199, 200, 201, 202, 203, 204, 205, 206, 207, 212, 216, 217, 218, 236
conservation tillage · 3, 5, 6, 7, 48, 52, 53, 54, 55, 56, 57, 98, 145, 164, 172, 181, 188, 190, 197, 202, 204, 205, 207, 212, 215, 217, 231, 232, 233, 234, 235, 236, 237, 240, 241, 243, 244, 246, 247, 248, 249, 250, 257, 258, 259, 260, 261, 262, 263,

264, 265, 270, 271, 272, 276, 277, 278, 279, 280, 281, 282, 283, 284
Conservation Tillage · 5, 52, 59, 60, 61, 63, 64, 104, 177, 191, 192, 193, 205, 216, 218, 219, 220, 221, 222, 223, 224, 226, 227, 231, 237, 241, 251, 252, 254, 255, 256, 257, 273, 274, 275, 286, 287, 288, 289, 290
Consumer Demand · 151
Contour Farming · 104, 208, 210, 213, 216, 218, 219, 220, 221, 222
corn · 5, 6, 7, 12, 13, 15, 16, 30, 31, 32, 34, 41, 47, 48, 49, 50, 51, 54, 55, 56, 75, 76, 78, 79, 80, 83, 92, 108, 111, 112, 136, 137, 138, 139, 142, 145, 148, 149, 150, 152, 182, 184, 186, 188, 189, 190, 191, 197, 206, 207, 209, 212, 215, 216, 217, 234, 236, 265, 267, 268, 270, 271, 272, 284, 285
cotton · 29, 30, 31, 32, 35, 41, 49, 50, 55, 78, 80, 83, 93, 108, 135, 139, 140, 142, 146, 150, 152, 188, 206
Critical Area Planting · 104
crop production · 17, 76, 77, 93, 138, 139, 145, 151, 161, 163, 164, 201
Crop Residue Management · 59, 62, 63, 64, 127, 177, 178, 179, 191, 192, 193, 225, 226, 251
Crop Residue Use · 104, 105
crop rotations · 3, 12, 17, 18, 31, 47, 48, 50, 51, 57, 93, 110, 113, 114, 150
Crop rotations · 12, 46, 48
cropland · 7, 12, 15, 47, 48, 52, 68, 76, 97, 160, 163, 165, 168, 169, 170, 184, 187, 188, 196, 200, 201, 202, 206, 215, 229, 234, 258, 259, 260, 262, 266, 268, 271, 272, 278, 279, 281, 282, 284, 285
Cropping Practices · 56, 182, 193, 208, 210, 213, 236, 238
crude oil · 6, 240, 241, 243, 246, 247, 248, 249, 250
cultivation · 17, 52, 53, 78, 112, 182, 188, 235
Cyanazine · 61, 253
Czech Republic · 121
CZMA · 71, 144

D

DDT · 18, 32, 41
deviation · 16, 239, 262, 263, 271
direct costs · 102
disease-resistance · 138
drinking water · 8, 19, 67, 72, 86, 171
durum wheat · 5, 189, 191, 206, 236

E

economic losses · 3, 9, 159, 185, 196
economic security · ix
economic variables · 252
Ecosystem · 150, 151, 154, 156
education · 85, 99, 100, 126, 172, 177, 205, 206, 223, 225
emission standards · 72
endangered species · 70, 72
environmental benefits · 5, 47, 56, 57, 77, 163, 166, 181, 188, 189, 190, 201, 203
environmental factors · ix
environmental impact · 54, 109, 257
environmental protection · 107, 109
environmental quality · 1, 31, 55, 57, 85, 87, 94, 98, 102, 108, 109, 110, 175
EQIP · 107, 163, 165, 200, 203, 212
erosion · ix, 2, 5, 6, 17, 46, 49, 52, 78, 107, 159, 160, 161, 162, 163, 164, 165, 166, 168, 169, 170, 171, 172, 173, 175, 181, 182, 183, 185, 187, 190, 195, 196, 198, 199, 201, 202, 205, 207, 212, 215, 217, 229, 231, 234, 235, 236, 257, 270, 278, 284
eutrophication · 2, 159, 185, 186, 195, 196, 197
exogenous factors · 234

F

FAIR · 4, 5, 110, 111, 112, 113, 114, 140, 149, 150, 165, 175, 202, 203
fallow land · 49
farm income · 5, 47, 50, 51, 52, 57, 91, 97, 109, 110, 145, 161, 175, 258, 259, 265
farm profitability · 46, 79, 95
farm workers · 45, 70, 109, 118, 119
farmers · 2, 4, 5, 9, 10, 14, 15, 16, 18, 20, 45, 48, 55, 56, 68, 73, 77, 85, 86, 87, 88, 91, 93, 94, 96, 97, 98, 100, 101, 102, 103, 107, 108, 111, 112, 138, 140, 144, 148, 149, 151, 159, 161, 162, 164, 165, 172, 173, 174, 175, 176, 182, 184, 185, 188, 189, 190, 191, 233, 234, 235, 236, 240, 257, 259, 264, 265, 268, 271, 280, 283
farming practices · 9, 45, 96, 100, 172
Farming Practices · 62
Fertilizer Application · 43, 61, 129
fertilizer residues · 8
Fertilizer Use · 12, 13, 44, 61, 62, 124, 125, 128, 183
FFDCA · 70, 71, 74, 79, 141, 146
FIFRA · 69, 71, 74, 141, 146, 147
financial assistance · 4, 5, 6, 86, 96, 99, 102, 103, 107, 172, 173, 176, 218
financial incentives · 95
financial support · 162, 199
flooding · 160, 186, 196
Florida · 89
Food Quality Protection Act · 8, 70, 71, 123, 141, 149, 154
food residues · 9
food safety · 1, 79
Food Security · 5, 46, 48, 164, 165, 168, 175, 201, 203, 234
Food, Agriculture, Conservation, and Trade Act · 46, 152
Forage · 61
FQPA · 8, 71, 141, 142, 143, 149, 154
fruits · 20, 31, 78, 82, 112, 137, 152
Fungicides · 33, 40

G

gasoline · 240, 267, 280
geography · 178, 225
Georgia · 89
goods and services · 150
government intervention · x, 67, 68, 69, 84
Government Policy · 5, 67, 133, 144, 146
government subsidies · x
Grading · 114, 115
Grain Crops · 61
Grain Yield · 62
Grazing · 105, 106
groundwater · 1, 17, 34, 50, 53, 69, 73, 86, 91, 94, 95, 97, 108, 118, 187, 197, 258
groundwater quality · 1, 59, 60, 62, 63, 64, 86, 120, 191, 192, 193, 223, 224, 226, 251, 252, 254, 256, 286, 287, 288, 289, 290

H

Habitat Management · 106
harvest · 110, 146, 181, 188
Hawaii · 89
Herbicides · 30, 40, 64, 186, 197, 255, 290
Host Plant Resistance · 138
House of Representatives · 111, 130
human health · 1, 2, 35, 36, 38, 41, 45, 67, 69, 70, 71, 77, 143, 144, 198

I

imports · 82
Indiana · 77, 89, 224
insect infestation · 3, 33
insecticides · 20, 21, 25, 26, 27, 28, 30, 31, 32, 41, 48, 49, 50, 55, 78, 80, 81, 139, 140, 145, 186, 197, 267
Integrated Crop Management · 107, 122, 127, 176
Internet · 162, 200, 205

Iowa · 50, 51, 61, 62, 64, 65, 89, 91, 93, 96, 124, 125, 129, 177, 179, 192, 193, 205, 224, 227, 286
irrigation · 92, 98, 145, 160, 171, 198, 280, 283
Irrigation · 43, 61, 104, 105

K

Kansas · 89, 120, 122, 176, 235, 256
Kentucky · 89, 234

L

labor costs · 54, 55, 182
labor force · 10
land retirement · 5, 6, 172, 174, 176, 218
Land Retirement · 174
Land Use · 61
landscaping · 160, 187
leaching · 11, 34, 48, 50, 51, 53, 69, 87, 88, 91, 93, 97, 108, 109, 187, 197, 215, 233
legislation · 5, 8, 73, 114, 142, 149, 152, 175, 236
Legumes · 12, 61, 63, 105
Levels · 72
lime · 89
Louisiana · 89, 100, 126

M

management skills · 6, 198, 204, 217, 234
market failure · 68, 87, 94
market forces · 117
market price signals · x
market prices · 2
mass media · 98, 233
Methyl bromide · 72
Miami · 177, 224
Microbial Pesticides · 134
mites · 138
monocropping · 47, 50, 51
Mulching · 105

N

national interest · 162, 199
National Research Council · 12, 17, 44, 47, 54, 55, 63, 71, 74, 76, 77, 111, 126, 146, 151, 155, 159, 178, 185, 193, 195, 198, 217, 225, 231, 254, 257, 279, 285, 289
natural resources · ix, 68, 150, 166, 203
new production practices · x, 133
nitrates · 53
nitrogen · 11, 12, 14, 15, 16, 17, 46, 48, 50, 51, 73, 76, 77, 89, 91, 92, 93, 94, 98, 101, 102, 108, 109, 118, 186, 197, 215
Nitrogen Cycle · 63, 128
Nitrogen Fertilizer · 43, 61, 62, 124
Nitrogen Loss · 43, 61
Nitrogen Testing · 59, 60, 120, 123, 176, 223
no till · 5, 7, 52, 54, 181, 182, 183, 184, 185, 186, 187, 188, 189, 190, 191, 197, 207, 212, 216, 217, 218, 219, 220, 221, 222, 235, 257, 258, 271, 272, 273, 276, 277, 278, 279, 280, 281, 282, 283, 284, 285
North Carolina · 90, 123, 155
North Dakota · 90
NRI · 166, 167, 170, 171, 266
Nutrient · 63, 104, 105, 106, 126, 127, 215, 224
nutrient management · 9, 55, 98, 101, 107, 145, 205

O

oats · 47, 142, 150
Organic Farming · 64
Organic foods · 151
ozone · 72

P

Pacific · 59, 137, 151, 205, 227

Pasture and Hayland · 104, 105
Pennsylvania · 90
Pest Management · 9, 43, 60, 104, 105, 121, 130, 137, 154, 155, 156, 176, 192, 252, 287
pest resistance · 9, 83, 139, 142, 151
Pest scouting · 145
pesticide applications · 7, 103, 108, 137, 145, 279, 280, 282, 285
pesticide contamination · 2, 185, 195
pesticide intensive farming · 133
pesticide spraying · 68
pesticides · 1, 2, 4, 7, 8, 9, 10, 11, 12, 18, 19, 20, 22, 26, 28, 30, 31, 33, 34, 35, 36, 37, 38, 39, 41, 42, 48, 49, 50, 53, 54, 56, 57, 58, 67, 68, 69, 71, 72, 74, 75, 77, 78, 79, 83, 87, 91, 93, 107, 109, 110, 114, 115, 116, 117, 133, 134, 138, 139, 140, 141, 142, 143, 144, 145, 146, 147, 148, 149, 150, 182, 186, 187, 196, 197, 215, 232, 233, 236, 265, 278, 279, 282, 283, 284
Pheromones · 134, 136
photosynthesis · 186, 197
pollution · 4, 57, 69, 72, 73, 85, 87, 88, 92, 94, 95, 96, 100, 102, 108, 144, 162, 175, 197, 200
pollution abatement · 4
population density · 171
Potassium · 209, 211, 214
Potatoes · 21, 22, 24, 27, 155
potential risks · 70, 74, 146
Prague · 121
Precision farming · 145
price of energy · 6, 237, 240, 243, 244, 246, 247, 248, 250
prices · ix, 2, 9, 14, 15, 18, 75, 78, 84, 92, 93, 109, 110, 111, 113, 114, 115, 116, 117, 140, 146, 150, 153, 240, 262, 266
productivity · 3, 4, 10, 11, 19, 54, 77, 113, 133, 144, 150, 159, 175, 184, 185, 196, 198, 202, 229, 235, 270, 271, 277, 279, 280, 281, 283, 284
public apprehension · 1
public interest · 174
Puerto Rico · 14, 90

Q

quality of the environment · ix

R

rainfall · 7, 52, 53, 228, 229, 231, 257, 268, 270, 272, 278, 281, 284
rate, survival · 48
recreation · 1, 151, 160, 171, 185, 186, 196
regulation · x, 3, 4, 5, 6, 58, 68, 69, 74, 75, 77, 83, 84, 85, 97, 101, 141, 144, 147, 148, 149, 172, 175, 176, 218
research and development · x, 5, 6, 74, 134, 138, 146, 147, 149, 172, 176, 201, 218
Resource Conservation and Recovery Act · 73
resource management · 1
Resource Protection · 176
resource scarcities · 1
Rhode Island · 90
Risk Analysis · 64
runoff · 2, 7, 48, 51, 52, 53, 69, 95, 97, 110, 159, 160, 185, 186, 187, 195, 196, 197, 198, 215, 228, 229, 231, 232, 233, 278, 283

S

Safe Drinking Water Act · 71, 72, 144
Safe Drinking Water Amendments Act · 9
salinization · ix, 195
sampling technique · 167
Seeding · 106
set-asides · 112, 149
Simazine · 60, 252, 288
social benefits · 94, 95, 163, 165, 171, 174, 200, 202
social costs · 5, 86, 171, 172, 176
soil degradation · ix, 2, 6, 195, 197
soil depth · 4, 175

soil erosion · ix, x, 3, 4, 7, 46, 52, 54, 56, 159, 160, 162, 163, 165, 166, 167, 170, 171, 172, 175, 176, 181, 184, 185, 187, 197, 198, 200, 202, 205, 206, 212, 215, 216, 229, 231, 233, 257, 258, 278, 283, 284, 285
soil loss · 46, 48, 185, 228, 229
soil moisture · 12, 30, 47, 53, 202
soil quality · 1, 17, 47, 164, 197, 198, 202
soybeans · 5, 6, 12, 30, 33, 35, 41, 47, 48, 49, 50, 51, 54, 55, 56, 75, 78, 79, 80, 142, 145, 148, 150, 184, 186, 188, 189, 191, 197, 206, 207, 211, 212, 215, 216, 217, 236
spring wheat · 5, 189, 191, 206, 236
standards · 9, 69, 71, 72, 74, 95, 113, 114, 115, 116, 117, 118, 141, 142, 147, 150, 152, 168
storm intensity · 278
Strip Crop-ping · 216, 218, 219, 220, 221, 222
surface water · 1, 17, 34, 53, 69, 73, 86, 108, 160, 185, 187, 196, 197, 232, 258
synthetic fertilizers · 46, 76

T

target price · 110
taxes · x, 4, 5, 6, 58, 68, 86, 91, 93, 94, 95, 172, 176, 218
technical assistance · x, 5, 6, 56, 57, 68, 85, 95, 99, 101, 103, 107, 108, 144, 161, 162, 163, 166, 172, 174, 176, 199, 200, 203, 204, 205, 206, 218
technological constraints · 3
Tennessee · 14, 90, 91
Till Farming · 5, 181
Tillage · 43, 52, 55, 59, 60, 61, 63, 64, 121, 176, 177, 182, 183, 188, 191, 192, 193, 194, 205, 208, 209, 210, 211, 213, 214, 217, 223, 224, 225, 226, 237, 246, 251, 252, 253, 254, 256, 269, 281, 286, 287, 288, 289, 290, 291
toxic dump sites · 73
toxic pollutants · ix, 195
Toxic Salt Reduction · 106
toxicity · 11, 19, 20, 35, 36, 37, 38, 39, 40, 41, 42, 69, 74, 109, 141, 147

U

U.S. Environmental Protection Agency, (EPA) · 34, 36, 43, 69, 70, 72, 74, 75, 122, 141, 142, 143, 147, 148, 149, 154
unemployment · 240
unemployment rate · 240
urban water supplies · 1
USDA · 43, 80, 83, 85, 110, 116, 152, 161, 162, 163, 164, 177, 179, 198, 199, 200, 201, 224, 287
Utah · 90

V

vegetables · 20, 31, 33, 35, 41, 78, 80, 82, 112, 137, 152
Voluntary Programs · 57, 63, 96, 108

W

Waste Management · 106
Waste Utilization · 104, 105, 106
water conservation · 162, 166, 200
water quality · ix, 2, 3, 5, 6, 46, 53, 73, 86, 96, 97, 102, 107, 108, 159, 161, 171, 185, 186, 190, 195, 196, 197, 198, 199, 206, 212, 216, 217, 231
Water Quality Program · 102, 107
Water quality regulations · 9
water resources · 46
water supplies · 72, 171
water supply · 96
water treatment · 186, 196
water users · 3, 159, 185, 196
Watersheds · 59, 60, 251, 252, 286, 288
weed management program · 183
weevils · 138

wildlife · 1, 19, 69, 70, 74, 118, 141, 147, 161, 163, 166, 200, 203
winter wheat · 5, 6, 30, 189, 191, 206, 207, 212, 214, 215, 216, 217, 236
Worker Protection Standards · 70
WQIP · 107, 108, 109